TRIPLES

Applications of Pythagorean Triples

India's Scientific Heritage

Founder Editor : Late Dr L M Singhvi

General Editor

Dr S N Bhavsar

Guest Editor

Dr Murli Manohar Joshi

5

Editorial Panel

TRIPLES

Applications of Pythagorean Triples

KENNETH WILLIAMS

Foreword by

L.M. SINGHVI

Formerly High Commissioner for India in the UK

MOTILAL BANARSIDASS PUBLISHERS
PRIVATE LIMITED • DELHI

5th Revised Edition: ***2009***
First Indian Edition: Delhi, 2003

(First Published by Inspiration Book in 1984)

ISBN: 978-81-208-1957-3 (Cloth)
ISBN: 978-81-208-1958-0 (Paper)

MOTILAL BANARSIDASS
41 U.A. Bungalow Road, Jawahar Nagar, Delhi 110 007
8 Mahalaxmi Chamber, 22 Bhulabhai Desai Road, Mumbai 400 026
236, 9th Main III Block, Jayanagar, Bangalore 560 011
203 Royapettah High Road, Mylapore, Chennai 600 004
Sanas Plaza, 1302 Baji Rao Road, Pune 411 002
8 Camac Street, Kolkata 700 017
Ashok Rajpath, Patna 800 004
Chowk, Varanasi 221 001

Printed in India
BY JAINENDRA PRAKASH JAIN AT SHRI JAINENDRA PRESS,
A-45 NARAINA, PHASE-I, NEW DELHI 110 028
AND PUBLISHED BY NARENDRA PRAKASH JAIN FOR
MOTILAL BANARSIDASS PUBLISHERS PRIVATE LIMITED,
BUNGALOW ROAD, DELHI 110 007

To

RUTH

PREFACE

The unchanging laws of number have always been a source of delight and inspiration. Particularly attractive are the Pythagorean triples which have so many elegant and interesting properties. But these triples are also of great practical use: through the theorem of Pythagoras the triples link the three main branches of mathematics: number, algebra and geometry.

This appears to be the first time that these triples have been developed into a useful structure, having applications in trigonometry, transformations in 2 and 3 dimensions, coordinate geometry in 2 and 3 dimensions, solution of triangles and equations, complex numbers, hyperbolic functions, simple harmonic motion, astronomy etc. Many more applications are also likely to appear. This book shows how the triples (and their 3-dimensional equivalent, quadruples) can be developed and applied and how they form a unifying thread linking many areas of mathematics.

Several chapters from previous editions of this book have been replaced or rewritten. Spherical trigonometry and prediction of planetary positions have gone and chapters on applied mathematics applications, the triple method and hyperbolic functions are now included. Topics which have been extended from previous editions include trigonometric equations and rotation of curves in 3-dimensional space. The chapter on angles in perfect triples has been improved and the chapter on sine, cosine, tangent and their inverses has been completely changed.

This book also serves as an illustration of Vedic Mathematics: a mathematical system which has been rediscovered by Sri Bharati Krsna Tirthaji (1884-1960) from ancient Vedic texts and is expounded in his book (see Reference 1). This system is based on sixteen formulae which are said to give one line answers to all mathematical problems. Being based on fundamental principles these Vedic formulae are therefore conspicuous in any structure that is developed in a simple and natural way. As the triples idea is introduced and extended in this book the operation of these formulae is evident. The formulae are expressed in word form (for example, *By One More than the One Before*) and as they arise in the text they are indicated by italic type. An index of these formulae will be found at the end of the book.

The diagram below gives a guide as to how the chapters in this book depend upon each other, so that Chapter 8 for example, can be understood by first reading only Chapters 1, 2 and 6.

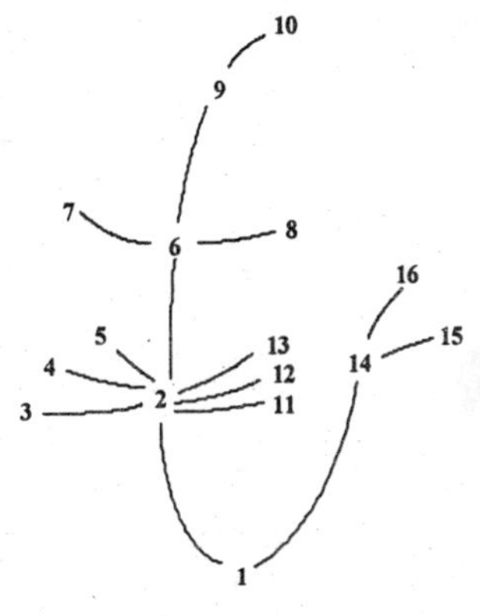

FOREWORD

Through trackless centuries of Indian history, Mathematics has always occupied the pride of place in India's scientific heritage. The poet aptly proclaims the primacy of the Science of Mathematics in a vivid metaphor:

यथा शिखा मयूराणां, नागानां मणयो यथा।
तद्वद् वेदांगशास्त्राणाम् गणितं मूर्धिन स्थितम्।।

—वेदांग ज्योतिष*

"Like the crest of the peacock, like the gem on the head of a snake, so is mathematics at the head of all knowledge. "

Mathematics is universally regarded as the science of all sciences and "the priestess of definiteness and clarity". J .f .Herbert acknowledges that "everything that the greatest minds of all times have accomplished towards the comprehension of forms by means of concepts is gathered into one great science, Mathematics". In India's intellectual history and no less in the intellectual history of other civilisations, Mathematics stands forth as that which unites and mediates between Man and Nature, inner and outer world, thought and perception.

Indian Mathematics belongs not only to an hoary antiquity but is a living discipline with a potential for manifold modem applications. It takes its inspiration from the pioneering, though unfinished work of the late Bharati Krishna Tirthaji, a former Sankaracharya of Puri of revered memory who reconstructed a unique system on the basis of ancient Indian tradition of mathematics. British teachers have prepared textbooks of Vedic Mathematics for British Schools... Vedic mathematics is thus a bridge across centuries, civilisations, linguistic barriers and national frontiers.

Vedic mathematics is not ont.1 a sophisticated pedagogic and research tool but also an introduction to an ancient civilisation. It takes us back to many millennia of India's mathematical heritage. Rooted in the ancient Vedic sources which heralded the dawn of human history and : illumined by their erudite exegesis, India's intellectual, scientific and aesthetic vitality blossomed and triumphed not only in philosophy, physics, astronomy, ecology and performing arts but also in geometry , algebra and arithmetic. Indian mathematicians gave the world the numerals now in universal use. The crowning glory of Indian mathematics was the invention of zero and the introduction of decimal notation without which mathematics as a scientific discipline could not have made much headway. It is noteworthy that the ancient Greeks and Romans did not have the decimal notation

---

* Lagadha, Verse 35

and, therefore, did not make much progress in the numerical sciences. The Arabs first learnt the decimal notation from Indians and introduced it into Europe. The renowned Arabic scholar, Alberuni or Abu Raihan, who was born in 973 A.D. and travelled to India, testified that the Indian attainments in mathematics were unrivalled and unsurpassed. In keeping with that ingrained tradition of mathematics in India, S: Ramanujan, "the man who knew infinity", the genius who was one of the greatest mathematicians of our time and the mystic for whom "a mathematical equation had a meaning because it expressed a thought of God", blazed new mathematical trails in Cambridge University in the second decade of the twentieth century even though he did not himself possess a university degree.

I do not wish to claim for Vedic Mathematics as we know it today the status of a discipline which has perfect answers to every problem. I do however question those who mindlessly deride the very idea and nomenclature of Vedic mathematics and regard it as an anathema. They are obviously affiliated to ideological prejudice and their ignorance is matched only by their arrogance. Their mindset were bequeathed to them by Macaulay who knew next to nothing of India's scientific and cultural heritage. They suffer from an incurable lack of self-esteem coupled with an irrational and obscurantist unwillingness to celebrate the glory of Indian achievements in the disciplines of mathematics, astronomy, architecture, town planning, physics, philosophy, metaphysics, metallurgy, botany and medicine. They are as conceited and dogmatic in rejecting Vedic Mathematics as those who naively attribute every single invention and discovery in human history to our ancestors of antiquity .Let us reinstate reasons as well as intuition and let us give a fair chance to the valuable insights of the past. Let us use that precious knowledge as a building block. To the detractors of Vedic Mathematics I would like to make a plea for sanity, objectivity and balance. They do not have to abuse or disown the past in order to praise the present.

Dr. L.M. Singhvi

Formerly High Commissioner for India in the UK

CONTENTS

THIS OLD BABYLONIAN TABLET DATING FROM 1900-1600 BC
LISTS FIFTEEN PYTHAGOREAN TRIPLES

1 TRIPLES

From unity the series of natural numbers can be generated *By One More than the One Before*:

1 2 3 4 5 6 7 8 9 . . . .

The series of perfect squares can then be formed:

	1	4	9	16	25	36	49	64	81	100	121	144	169 . . .
(that is	1^2	2^2	3^2	4^2	5^2	6^2	7^2	8^2	9^2	10^2	11^2	12^2	13^2 . . . .)

Then we may observe that the "triples" 3, 4, 5
6, 8, 10
5, 12, 13

each have the property that the sum of the squares of the first two numbers is equal to the square of the third.
For example, for 3, 4, 5 we have $3^2 + 4^2 = 5^2$ because $9 + 16 = 25$.
Similarly for 6, 8, 10 and 5, 12, 13.
Any three numbers having this property will be called a **triple**.

Other examples are:

9	12	15
8	15	17
1.5	2	2.5
–4	3	5
2	3	$\sqrt{13}$

Triples in which all the three parts (or **elements**) are rational numbers (that is, whole numbers or fractions, positive or negative) we will call **perfect triples**.
All the triples above are perfect except 2, 3, $\sqrt{13}$ since the square root of 13 is irrational.
The number of perfect triples is infinite.
We will see later how to generate perfect triples and how to generate triples with certain specified properties.

THE TRIPLE THEOREM

Take a square of side 7 units and join the points on the sides 3 units from each corner, as shown:

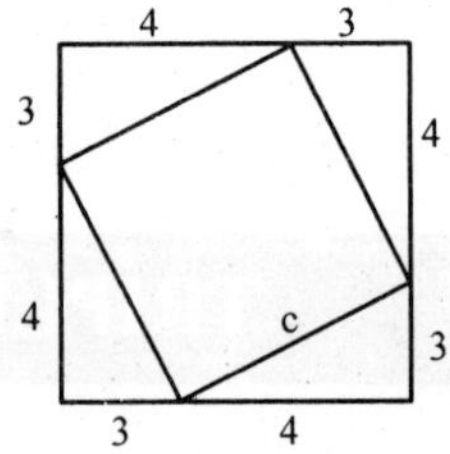

The resulting figure is also a square. Let c be the side of this square.
Then since: **area of smaller square + area of 4 triangles = area of larger square,**

$$\mathbf{c^2} \qquad + \qquad \mathbf{4 \times \tfrac{4\times3}{2}} \qquad = \qquad \mathbf{7^2}$$

Therefore $c^2 + 24 = 49$ and $c^2 = 25$. So $\underline{c = 5}$.

This means that if a triangle is formed with sides equal to the elements of the perfect triple 4,3,5 it will be a right-angled triangle.

Similarly, if the larger square has sides a+b, as shown:

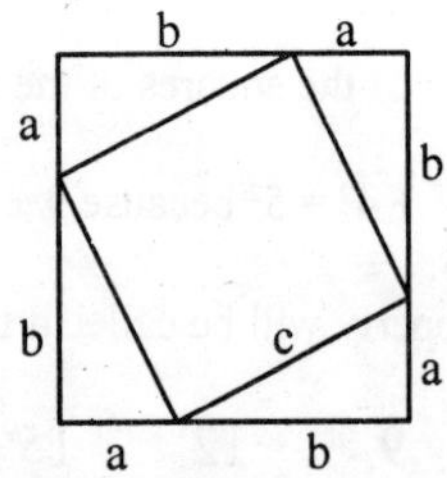

then $c^2 + 4 \times \frac{ab}{2} = (a + b)^2$
$\therefore\ c^2 + 2ab = a^2 + 2ab + b^2$
$\therefore\ c^2 = a^2 + b^2$
That is: a, b, c is a triple.

So the lengths of the sides of any right-angled triangle form a triple: this is the triple theorem (and it is practically the same as Pythagoras' theorem).

Conversely, if three numbers form a triple they represent the sides of a right-angled triangle.

The Vedic formula under which this important theorem comes is *The Sum of the Products is the Product of the Sum*, where products means squares in this case and the hypotenuse (longest side) is considered as the sum of the other two sides.

Example 1

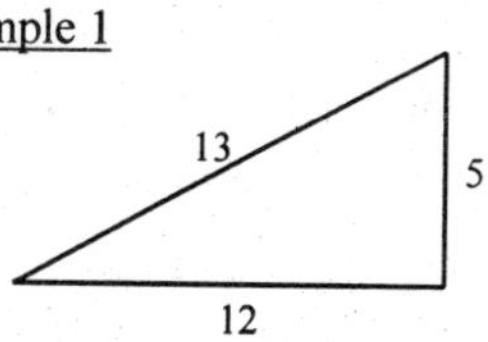

Since $12^2 + 13^2 = 13^2$ $(144 + 25 = 169)$
It follows that 12, 5, 13 is a triple.
This means that the triangle must be right-angled.

Example 2

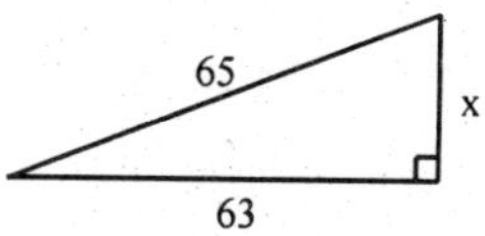

We have a right-angled triangle, so
63, x, 65 is a triple.
Therefore $x^2 = 65^2 - 63^2$.
By Addition and by Subtraction we get
$x^2 = (65+63)(65-63) = 128 \times 2 = 256$.
Therefore $\underline{x = 16}$.

Note that no attempt is generally made to draw the diagrams accurately.

SOME HISTORICAL BACKGROUND

There is considerable evidence that perfect triples have been known and used for many thousands of years. Many ancient stone circles, ellipses and egg-shaped structures are based on perfect triples, or triangles which very nearly satisfy Pythagoras' theorem, such as 7, 11, 13 (see References 2 and 3). The dimensions of altars in ancient India were based on perfect triples (see reference 4). The Old Babylonian Plimpton 322 tablet shown on Page (vi) is from the Plimpton Collection of Columbia University in New York and dates from 1900-1600 B.C. It is incomplete: the left-hand part is missing and the reverse is ruled with the same pattern of columns as the front but has no writing. The tablet lists 15 perfect triples, the angles in which begin at just under 45° and decrease in steps of 1° to about 32° (see References 5 and 6).

The theorem of Pythagoras was known long before Pythagoras (who lived c540 B.C.), the earliest statement of it being in the Indian Sulbasutras dated c800 B.C. (see Reference 4 Pages 14-22). For example, to quote from the Katyanana Sulbasutra "The cord in the diagonal of a square is the cord (the line) producing the double (area)". "The cord stretched in the diagonal of an oblong produces both (areas) which the cords forming the longer and the shorter side of an oblong produce separately" (see Reference 7). That is: the square of the diagonal of an oblong is equal to the square of both its sides.

NOTATION FOR TRIPLES

That a triple can represent the sides of a right-angled triangle is extremely useful and we will use the notation

A) 4, 3, 5

to represent at once a triple, a triangle and an angle. This can be read as: A is the angle contained in the triple 4, 3, 5, between the sides of length 4 and 5.

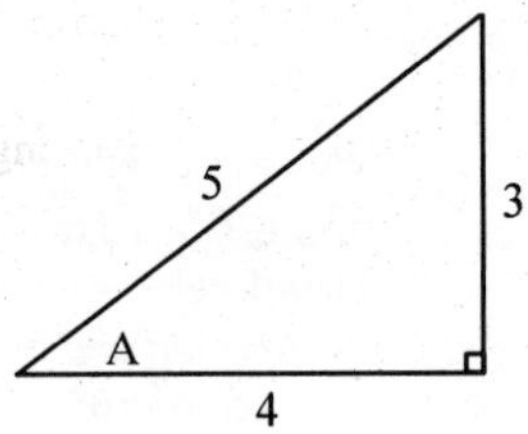

The angle, A, is put first, then the base, height and hypotenuse of the triangle in that order, the angle being between the base and the hypotenuse.
But we will use 4, 3, 5 etc. (without the angle) for a triple in which the angle is not important or where the angle or triangle is obvious.

With this notation all the trigonometric ratios are immediately available.
E.g. from A) 4, 3, 5 we have $\cos A = \frac{4}{5}$ etc.
Also, if the angle is at the origin of a rectangular coordinate system and the base of the triangle is along the positive x-axis then the first two elements of the triple, (4,3) in the above example, are the coordinates of the other end of the hypotenuse.
In this way the notation combines the cartesian and polar methods of describing the position of a point.

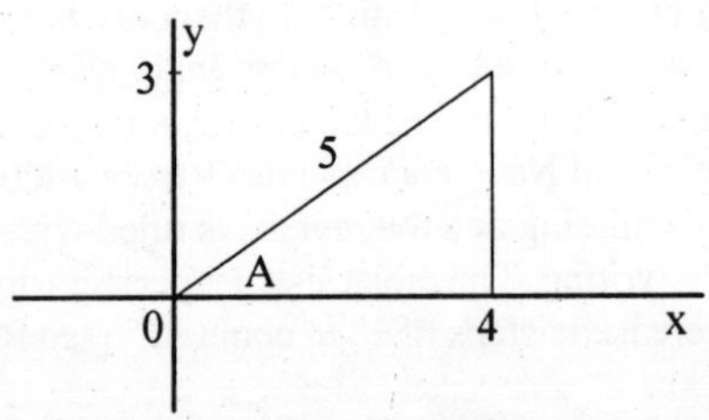

EQUAL, PRIME AND COMPLEMENTARY TRIPLES

If one triple is a fraction or multiple of another the triples will be called **equal triples.**
For example 4, 3, 5 = 8, 6, 10 = 2, 1½, 2½ etc.
This is because the triangles these triples represent all have the same shape (same angles).

A perfect triple is a **prime triple** if its elements are whole numbers and have no common factor other than 1.
So in every family of equal, perfect triples there is just one prime triple.
For the equal triples above the prime triple is 4, 3, 5.

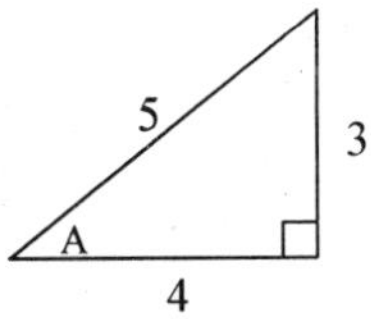

4, 3, 5

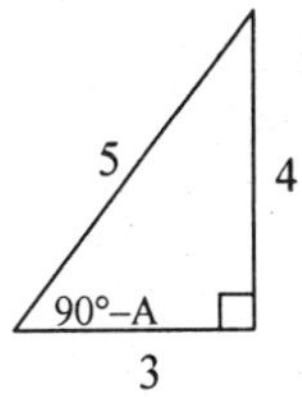

3, 4, 5

The triple 3, 4, 5 is the **complementary triple** to the triple 4, 3, 5 as it represents the complementary angle (it represents the third angle, i.e. the angle at the top of the 4, 3, 5 triangle shown above).
So to get the triple complementary to a given triple we *transpose* its first two elements.

SOME PERFECT TRIPLES

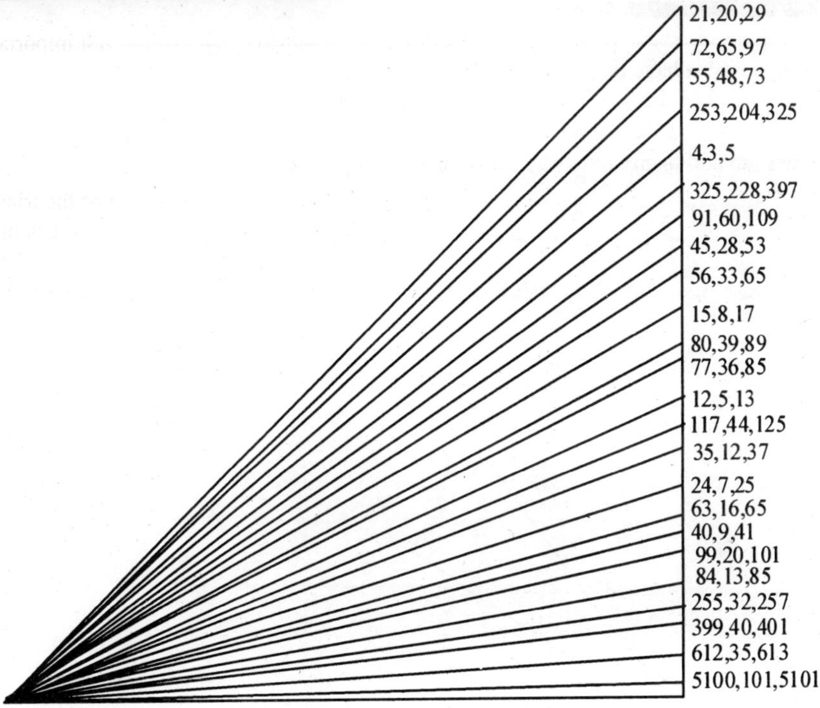

These 24 perfect triples are all constructed on the same base, the position of the triple numbers indicating the third vertex of each triangle.

2 TRIPLE ARITHMETIC

To develop a triple arithmetic we need some way of combining triples. In this chapter triple addition is defined; leading to triple subtraction and the doubling and halving of triples. Applications are shown in subsequent chapters and in Chapter 6 we develop this triple arithmetic further.

ADDITION OF TRIPLES

Example 1 Suppose we have two triples: A) 4 3 5
B) 15 8 17

and that we wish to add them in the way shown below, to produce a triangle OPQ containing the angle (A+B).

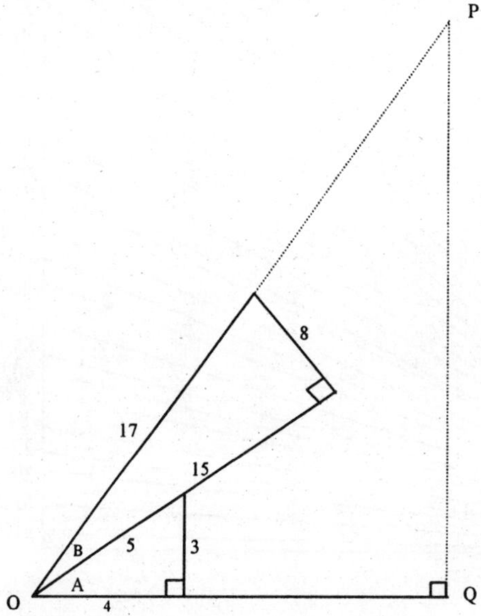

The triple which contains the angle (A+B) is given by:

A	4	3	5.
B	15	8	17
A+B	$(4 \times 15 - 3 \times 8)$,	$(3 \times 15 + 4 \times 8)$,	(5×17)
=	36	77	85

That is, we multiply vertically in the first two columns and subtract to get the first element of the triple for (A+B).
Then we cross-multiply in the first two columns and add to get the second element.
And we multiply vertically in the third column to get the third element.

These *Vertical and Crosswise* products are shown diagrammatically below:

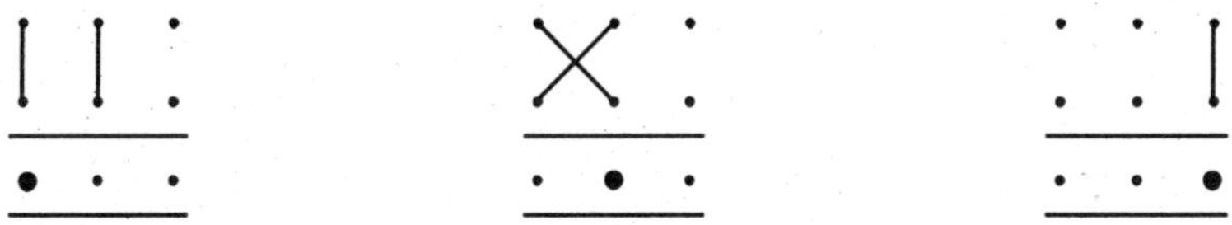

1st element 2nd element 3rd element

In general, for two triples A) x y r and B) X Y R the triple containing the angle (A+B) is given by:

A	x	y	r
B	X	Y	R
A+B	$(xX - yY)$,	$(yX + xY)$,	(rR)

This is derived on Page 160.

Thus we see that the sum of two perfect triples is itself a perfect triple!

It should be noted that the triple produced can always be divided or multiplied by any desired number without changing the shape of the triple.
If, for example, the hypotenuse of the triple 36, 77, 85 was required to be 17 (the hypotenuse of the added triangle) the triple could be divided through by 5 to give $\frac{36}{5}, \frac{77}{5}, 17$

<u>Example 2</u>

A	12	5	13
B	3	4	5 +
A+B	16	63	65

using the vertical and crosswise pattern shown above

Example 3

A	4	3	5	
B	24	7	25	+
A+B	75	100	125	any common factor
=	3	4	5	can be divided out

Example 4

A	24	7	25	
B	8	15	17	+
A+B	87	416	425	

two methods of simplifying this addition process, (useful when the triples have large numbers) are given later

DOUBLE ANGLE

Example 5

A	12	5	13	
A	12	5	13	+
2A	119	120	169	

It will be seen from this example that the procedure simplifies when finding the triple for a double angle:

A	12	5	13
2A	$(12^2 - 5^2)$,	$(2\times12\times5)$,	(13^2)

And in general:

A	x	y	r
2A	$(x^2 - y^2)$,	$(2xy)$,	(r^2)

Example 6

A	3	4	5
2A	–7	24	25

Here we find that the first element is negative. We can interpret by considering the addition of two 3,4,5 triangles:

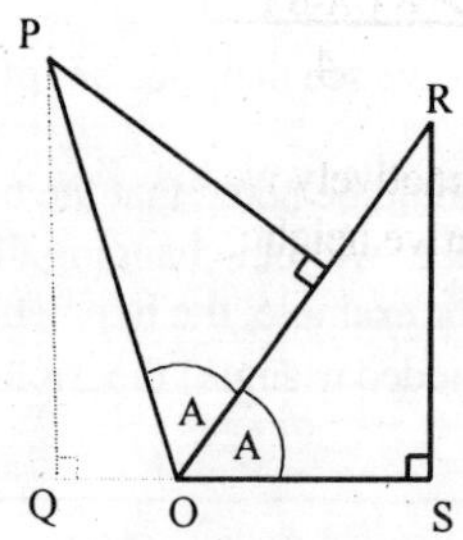

The sum of the two angles is obtuse and the resultant triangle OPQ has its base, OQ, extended in the opposite direction to the base OS.

So the triple X) –7, 24, 25 looks like this:

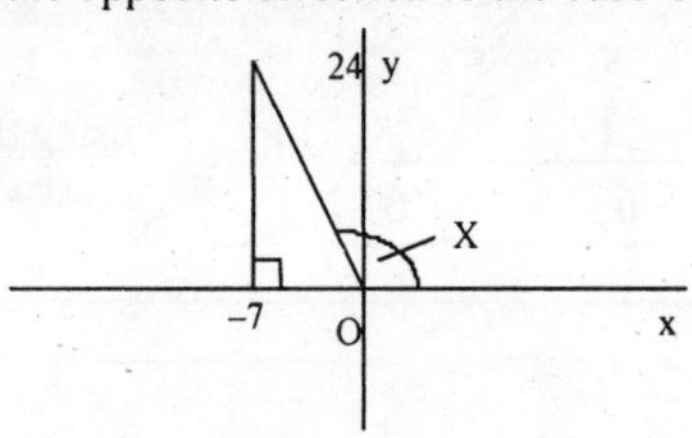

This is consistent with the familiar cartesian coordinate system with O the origin and OS along the x-axis and angles measured anticlockwise from the positive x-axis.
Thus if the first element of a triple is negative, and the second positive, the angle is obtuse.
And conversely, if the angle is obtuse the first two elements are negative and positive resp.

TRIPLE SUBTRACTION

Example 7 If A) 4 3 5 and B) 15 8 17 find a triple for A–B.

A	4	3	5
B	15	8	17
A–B	$(4 \times 15 + 3 \times 8)$,	$(3 \times 15 - 4 \times 8)$,	(5×17)
=	84	13	85

The diagram below shows how the angles are subtracted: the base of 15, 8, 17 is placed as usual on the hypotenuse of 4, 3, 5. But its hypotenuse is below the base so that the angle B is subtracted from A.

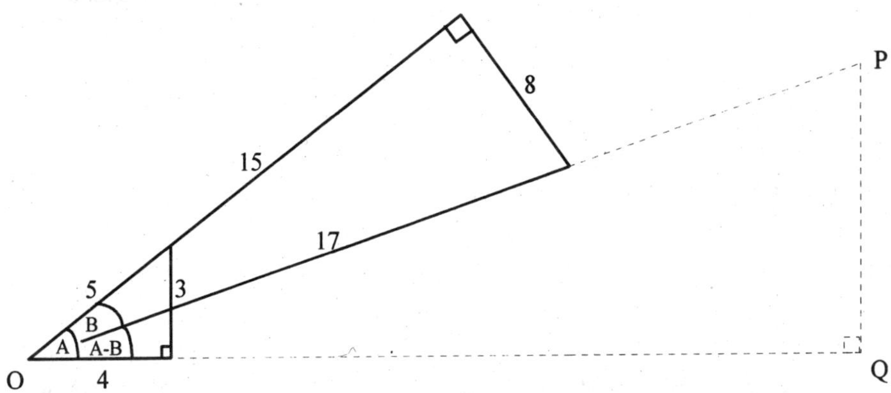

Alternatively we may consider that subtracting a triple is equivalent to adding a triple with a negative height:

A	4	3	5	
B	15	–8	17	–
A–B	84	13	85	as before.

Example 8

3	4	5	
4	3	5	–
24	7	25	

<u>Example 9</u>

A	4	3	5
B	3	4	5 −
A–B	24	−7	25

Here we get a triple with a negative height. This is because B, the angle being subtracted, is larger than the angle A:

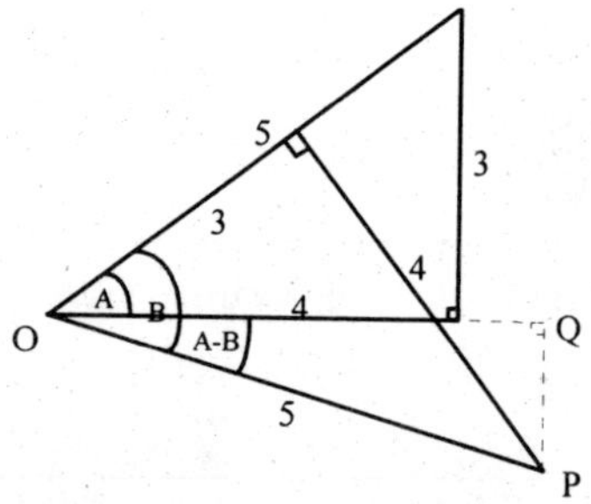

In fact a single triple can be used to represent Eight different angles: one in each quadrant.

For example:

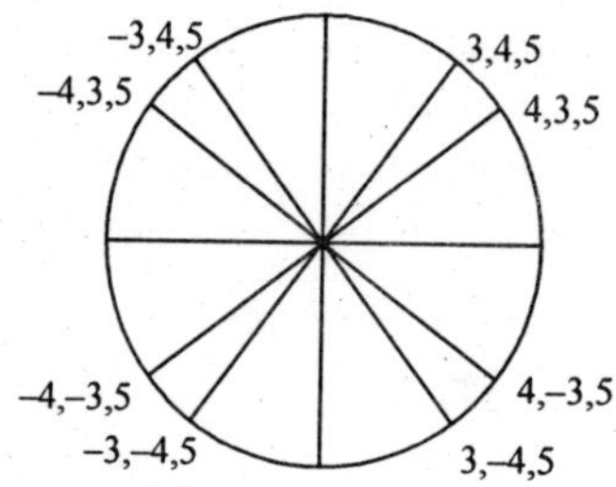

<u>Example 10</u> If A) −3 4 5 and B) 12 5 13 find a triple for 2A–B.

A	−3	4	5
2A	−7	−24	25
B	12	5	13 −
2A–B	−204	−253	325

Here 2A and 2A–B are both in the third quadrant.

Exercise A Add the following triples:

1.	2.	3.	4.	5.
3 4 5	24 7 25	40 9 41	12 5 13	8 15 17
15 8 17 +	3 4 5 +	4 3 5 +	8 15 17 +	4 3 5 +

Find a triple for 2A:

6. A)4, 3, 5 7. A)15, 8, 17 8. A) 21, 20, 29 9. A) 5, 12, 13 10. A)8, 15,17

Subtract the following triples:

11.	12.	13.	14.	15.
7 24 25	4 3 5	12 5 13	12 5 13	3 –4 5
3 4 5 –	12 5 13 –	24 7 25 –	3 4 5 –	3 4 5 –
______	______	______	______	______

Given A)12,5,13 B)5, 12, 13 C)3,–4,5 find triples for:

16. A–B 17. B–C 18. 2C–B 19. A–B–C 20. A–A

QUADRANT ANGLES

The quadrant angles, 0°, 90°, 180°, 270° can also be expressed by triples.

Example 11 Suppose we add the triples A)4 3 5 and B)3 4 5:

A	4	3	5
B	3	4	5 +
A+B	0	25	25
=	0	1	1

In this case we get a triangle with no base and height and hypotenuse equal.

We can divide 0,25,25 by 25 to get 0,1,1.

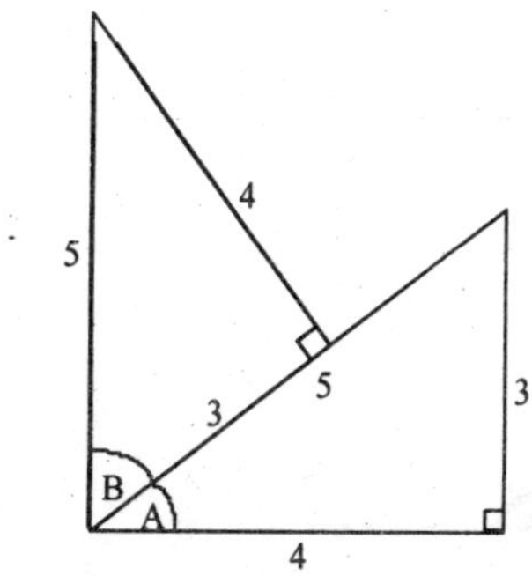

We see that because the triples are complementary A+B = 90°, so that the triple 0, 1, 1 represents an angle of 90°:

90°) 0 1 1

And if 90°) 0 1 1 we can double this triple to get a triple for an angle of 180°:

90°	0	1	1
180°	–1	0	1

Next we can double the 180° triple to get a triple for 360°, which is also a triple for 0°:

180°	–1	0	1
360°	1	0	1

And for 270° we can add 180° and 90°:

180°	−1	0	1
90°	0	1	1
270°	0	−1	1

These results can be summarized:

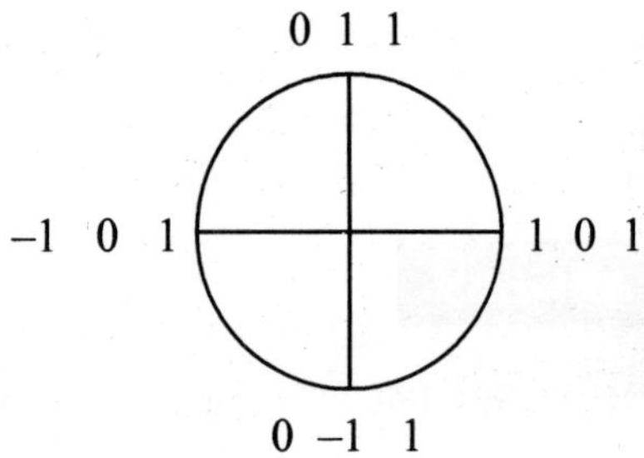

We can now use these triples in combination with others.

Example 12 Given A) 4 3 5 find a triple for A+90°.

A	4	3	5	
90°	0	1	1	+
A+90°	-3	4	5	

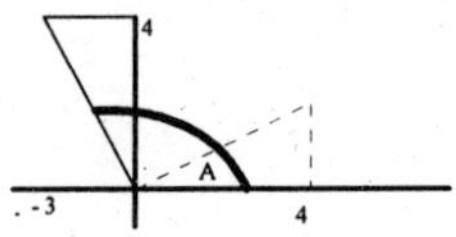

Example 13 Given A) 4 3 5 find a triple for 180°– A.

180°	-1	0	1	
A	4	3	5	-
180°-A	-4	3	5	

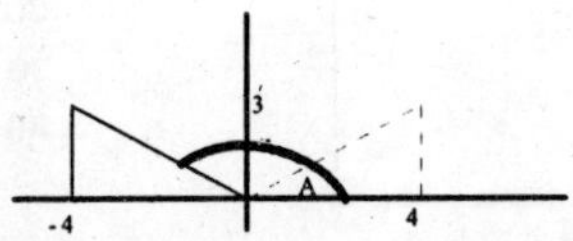

That is, we simply change the sign of the first element of a triple to obtain the supplementary triple.

Example 14 Given A) 4 3 5 find a triple for A+180°.

A	4	3	5	
180°	-1	0	1	+
A+180°	-4	-3	5	

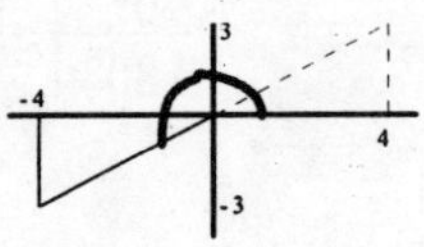

That is, we reverse the signs of the first two elements when adding 180°.

Example 15 Given A) $-2, -3, \sqrt{13}$ find a triple for $-A$.

0°	1	0	1
A	−2	−3	$\sqrt{13}$ −
−A	2	3	—

[We may insert a dash for an element of a triple that is determined by other elements or if it is unknown.]
In general we change the sign of the middle element of a triple A to obtain the triple for −A.
Geometrically the triangle is simply reflected in the horizontal axis.

TRIPLE GEOMETRY

We can solve triangles etc. by working with triples rather than angles given in degrees or radians.

Example 16 Two angles of a triangle are given by the triples 4, 3, 5 and 5, 12, 13.
Find the triple for the third angle.

Given two angles of a triangle we would normally add them and take the result from 180°.
Here we do the same thing with the triples:

4	3	5	
5	12	13	+
−16	63	65	

We can subtract this result from the triple for 180° or use the short cut suggested in Example 13 to get 16, 63, 65 for the answer.

Later we will see how to get this answer more easily.

ANGLES OF 30°, 60°, 45° ETC.

Triples containing these angles are easily found.

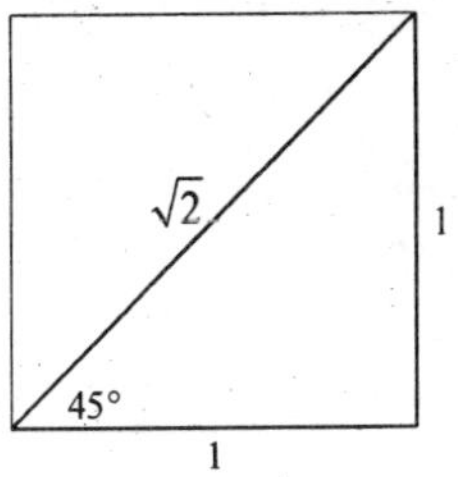

From half a square we get: 45°) 1 1 $\sqrt{2}$
Adding 90° to this we get: 135°) −1 1 $\sqrt{2}$

And from half an equilateral triangle of side 2 we get:

30° :	$\sqrt{3}$	1	2
60° :	1	$\sqrt{3}$	2
120° :	–1	$\sqrt{3}$	2
240° :	–1	$-\sqrt{3}$	2

etc.

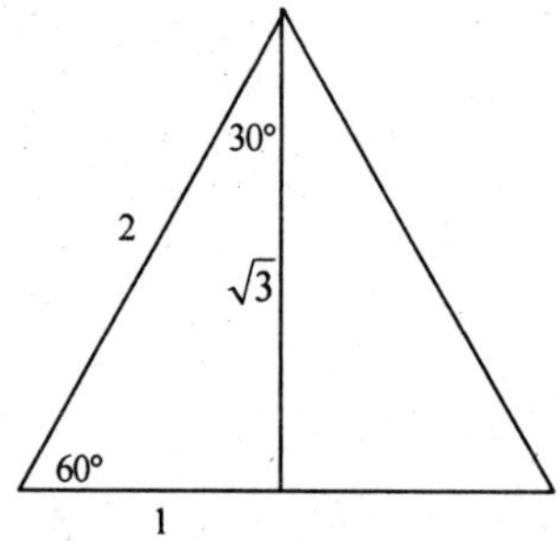

<u>Example 17</u> Find a triple for 75°.

75 = 30 + 45 so:

30°	$\sqrt{3}$	1	2	
45°	1	1	$\sqrt{2}$	+
75°	$\sqrt{3}-1$	$1+\sqrt{3}$	$2\sqrt{2}$	

Subtraction of a triple from the 45° triple and addition to the 45° triple are particularly easy, as the following examples show.
This illustrates another application of the formula *By Addition and By Subtraction.*

<u>Example 18</u> Subtract 12, 5 13 from the triple for 45°.

1	1	$\sqrt{2}$	
12	5	13	–
17	7	-	

That is, we just add 12 and 5 and then subtract them.

<u>Example 19</u> Add 45° to the angle in 12, 5, 13.

1	1	$\sqrt{2}$	
12	5	13	+
7	17	–	

Here we subtract 5 from 12 and then add them.

Exercise B Find the following triples:

1. Given A) 12, 5, 13 find A–90°
2. Given A) 3, 4, 5 find 270°+ A
3. Given A) 15, 8, 17 find A–180°
4. Given A) 4, 3, 5 find A+45°
5. Given A) 5, 12, 13 Find triples for
   a) –A
   b) 45°– A
   c) 2A+ 90°
   d) 270°– 2A

Find triples for:

6. 105°   7. 165°   8. –15°   9. 285°   10. 225°

11. Given that in triangle ABC, AB=AC and that a triple for B is B)12, 5, 13 show that a triple for angle BAC is –119, 120, 169.

HALF ANGLE

Example 20 Given A) 7 24 25 find a triple for ½A.

½A) 7+25, 24, - = 32, 24, - = 4, 3, 5

That is, we add the first and last elements of the triple to get the first element of the half-angle triple and keep the middle element as the middle element of the half-angle triple.
The third element can be calculated from the first two.

Proof:

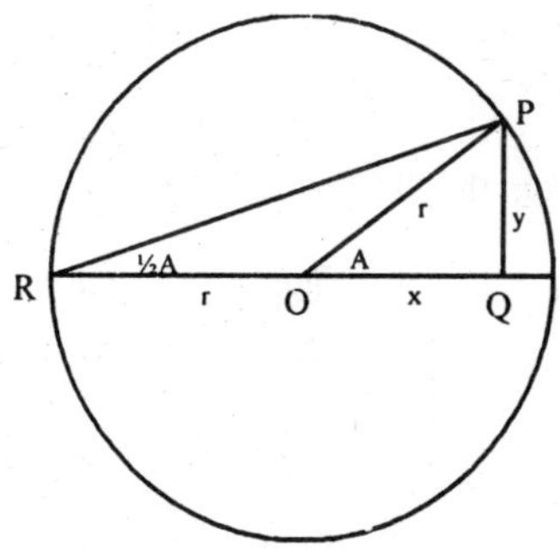

Given a triple A) x, y, r draw a circle centre O, radius r, as shown.
Produce QO to R and join RP.
Then the angle at R is ½A so that from ΔRPQ the triple for ½A is ½A) x+r, y, -.

Similarly if A is obtuse, so that x is negative:

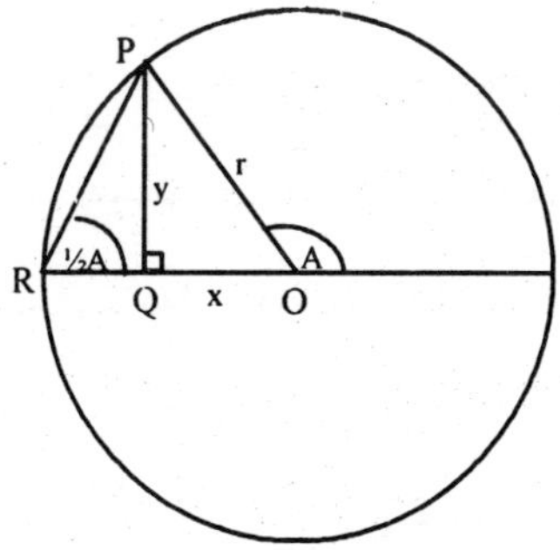

We still have ½A) x+r, y, -

Example 21 If A) 4, 3, 5 then ½A) 4+5, 3, - = 3, 1, $\sqrt{10}$.

Example 22 If A) –3, 4, 5 then ½A) 2, 4, - = 1, 2, $\sqrt{5}$.

In fact, because A and A+360° are represented by the same triple we generally have two solutions for a half angle triple in a 360° range : ½A and ½A+180°.
So in Example 21 in which we found ½A) 3, 1, $\sqrt{10}$ we know that –3, –1, $\sqrt{10}$ is also an answer (see Example 14 for addition of a 180° triple).

<u>Example 23</u> Given that A) –3, –4, 5 then ½A) 1, –2, $\sqrt{5}$
or ½A) –1, 2, $\sqrt{5}$

<u>Example 24</u>

In the diagram the angle A is given by The 3, 4, 5 triple.

Find a triple for the obtuse angle at C.

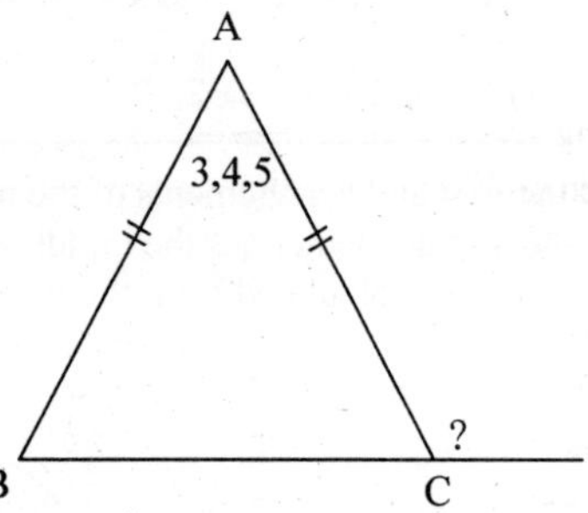

We could subtract 3,4,5 from the 180° triple, halve the result and then subtract from 180°.

180°	–1	0	1	
A	3	4	5	–
180°–A	–3	4	5	
½(180°–A)	1	2	$\sqrt{5}$	= B = C
180°–C	–1	2	$\sqrt{5}$	Answer

Alternatively we halve the isosceles triangle as shown, halve the 3,4,5 triple to get 2, 1, -, for the angle CAD,

find the complementary triple 1, 2, -, for the angle ACD

and then the supplementary triple <u>–1, 2, -</u>

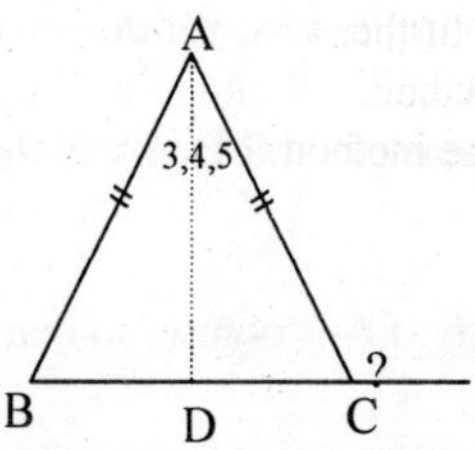

✵ Find triples for 15° and for 22½°. [$\sqrt{3}+2$, 1,– and $1+\sqrt{2}$,1,–]

Exercise C Find a half angle triple for:

1. 5,12,13 2. 15,8,17 3. 8,15,17 4. 119,120,169 5. –4,3,5

6. –119,120,169 7. –7,24,25 8. 24,–7,25

9. Given A) 3,4,5 find triples for a) 90°– A b) 2A– 90° c) ½A
d) ½A+A e) 4A f) 3A

SIMPLIFYING CALCULATIONS

Where large numbers are concerned this addition (and subtraction) process can be simplified in various ways.
In Example 4 earlier we found the first element by subtracting the vertical products.
Any common factor in either or both diagonals can be taken out.
In this case we see that 24 and 15 have a common factor of 3 and we can take this out so that the first two columns become

8	7
8	5

Multiplying vertically and then subtracting then gives 64–35=29.
This is then multiplied by the 3 that was taken out to give 87 as the first element.

Similarly when carrying out the cross-product to get the second element **any common factor in either or both verticals can be taken out.**
In this case there is a common factor of 8 in the first column and taking this out the first two columns become

3	7
1	15

Multiplying crosswise and adding then gives 7+45=52 and this 52 is then multiplied by the 8 that was taken out to give 416 as the second element of the answer.

Just the same method can be used for subtraction. In getting the first element in Example 7 we can take out the common factor of 3 in one of the diagonals.

A further simplifying device is given on Page 58 and this can be combined with the above method.
The methods of simplifying determinants can also be used here.

Right-angled triangles are a basic unit in geometry and by working with the perfect triples we can make use of their special properties and solve problems in a wide variety of situations.
The development of triple arithmetic in this chapter has been simple and natural and will be extended later.

The main results of the first two chapters are summarised below.
In the next three chapters we will see how varied the applications of these results are and how easily we can solve problems that normally require special formulae of their own.

SUMMARY

Given A) x y r — **Page**

1. x, y and r represent the sides of a triangle with base of length x, height y and hypotenuse of length r. A is the angle between the base and the hypotenuse.
2. Then $x^2+y^2=r^2$. — 2
3. A)x y r is called a **triple**.
4. If the elements of x,y,r are whole numbers with no common divisor then x,y,r is a **prime triple.** — 5
5. x,y,r and nx,ny,nr (where n is a real number) are **equal triples.** — 5
6. If x, y and r are rational numbers then x,y,r is a **perfect triple.** — 1
7. For **addition** of triples: — 6

$$\begin{array}{ccc} x & y & r \\ X & Y & R \ + \\ \hline xX-yY & yX+xY & rR \end{array}$$

8. For **subtraction** of triples: — 9

$$\begin{array}{ccc} x & y & r \\ X & Y & R \ - \\ \hline xX+yY & yX-xY & rR \end{array}$$

9. If A)x, y, r then 2A)x^2cy^2, 2xy, r^2 (**double** angle) — 8
10. If A)x, y, r then ½A) x+r, y, - (**half** angle)
    or ½A) –(x+r), –y, - — 15

3 TRIPLE TRIGONOMETRY

INTRODUCTION

Trigonometry, as the name suggests, is the study of the relations between the sides and angles of triangles. In this chapter we consider right-angled triangles in terms of triples. Non right-angled triangles are dealt with in a later chapter.

You will recall that when a triangle, say a 4,3,5 triangle, is in the position shown below (with A and the right angle at the bottom) then the ratio $\frac{\text{height}}{\text{hypotenuse}}$ is called the sine of angle A, abbreviated sinA.

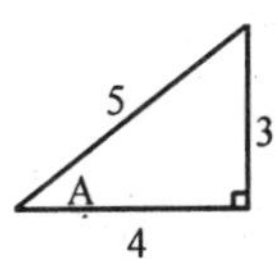

Thus we have sinA = $\frac{3}{5}$ in this case. This is a definition of sinA and since there are six possible ratios which can be formed from the three sides there are six definitions altogether.

The words sine, cosine, tangent, cosecant, secant, cotangent are abbreviated to sin, cos, tan, cosec, sec, cot

and the six definitions , referring to the triangle above, are:

$\sin A = \frac{3}{5}$ $\quad \cos A = \frac{4}{5}$ $\quad \tan A = \frac{3}{4}$ $\quad \text{cosec} A = \frac{5}{3}$ $\quad \sec A = \frac{5}{4}$ $\quad \cot A = \frac{4}{3}$

Thus cosA simply means the base of a triangle divided by the hypotenuse, and so on.

Note that these definitions will serve for angles of any size so that if, for example, A is obtuse:

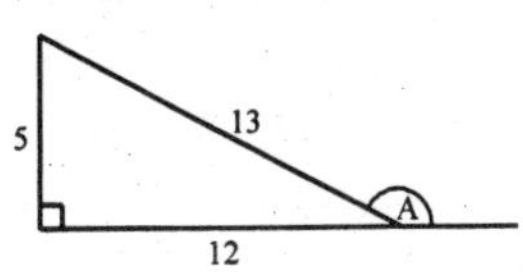

then $\sin A = \frac{5}{13}$ and $\cos A = -\frac{12}{13}$ because the base is extended in the negative direction for obtuse angles.

<u>Example 1</u> If $\sin A = \frac{3}{5}$ and $\cos B = \frac{12}{13}$, where both A and B are acute angles, find $\tan(A+B)$.

We have

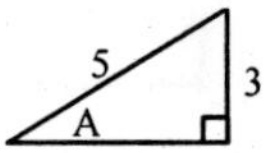

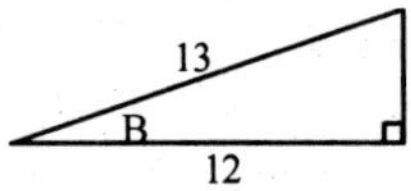

The third element of each triple is obtained from the triple theorem.
Adding these triples:

A	4	3	5	
B	12	5	13	+
A+B	33	56	65	

i.e.

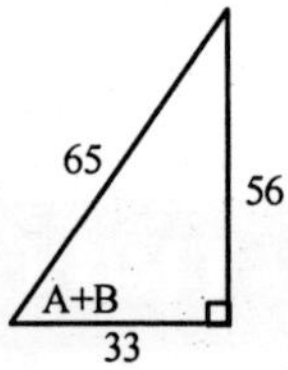

$\therefore \tan(A+B) = \frac{56}{33}$

If we were not told, in the above example, that the angles were both acute then the first element of the triple for A could be ± 4. Similarly the second element for B could be ± 5. This means there would be four cases to consider and four possible answers for $\tan(A+B)$.

<u>Example 2</u> If $\cos A = \frac{2}{3}$ and $\sin B = \frac{5}{6}$ find $\sin(A+B)$.

A	2	$\pm\sqrt{5}$	3	
B	$\pm\sqrt{11}$	5	6	+
A+B	–	$\pm\sqrt{55}+10$	18	

$\therefore \sin(A+B) = \frac{\pm\sqrt{55}+10}{18}$.

There is no need to find the first element of the triple for A+B.

<u>Example 3</u> If $\cos A = \frac{7}{25}$ and $\sin B = \frac{4}{5}$, where A is acute and B is obtuse, find $\cos(A-B)$.

A	7	24	25	
B	–3	4	5	–
A–B	75	-	125	

$\therefore \cos(A-B) = \frac{75}{125} = \frac{3}{5}$.

For obtuse angles the first element is negative and the second positive.

<u>Example 4</u> Given $\sin A = \frac{3}{5}$ and that A is acute, find $\tan(90°- A)$.

90°	0	1	1	
A	4	3	5	–
90°– A	3	4	5	

$\therefore \tan(90°- A) = \frac{4}{3}$.

<u>Example 5</u> Given A) -2, -3, $\sqrt{13}$ find cot(-A).

0°	1	0	1	
A	–2	–3	-	–
–A	–2	3	-	

$\therefore \cot(-A) = -\frac{2}{3}$.

Example 6 If $\tan A = \frac{24}{7}$ and A is acute, find sec(A–30°).

A	7	24	25	
30°	$\sqrt{3}$	1	2	–
A – 30°	$7\sqrt{3}+34$	-	50	

$\therefore$ $\sec(A–30°) = \frac{50}{7\sqrt{3}+24}$.

Example 7 If $\cot A = \frac{4}{3}$ find cos2A and tan½A (A is acute).

A	4	3	5
2A	7	24	25
½A	9	3	-

$\therefore$ $\cos 2A = \frac{7}{25}$ and $\tan\frac{1}{2}A = \frac{1}{3}$

Exercise A If $\sin A = \frac{4}{5}$ and $\cos B = \frac{12}{13}$ (A and B both acute) find:

1. sin(A+B) 2. cosec(B–A) 3. tan2A 4. cot(B+45°) 5. 2tan½B

If $\tan A = \frac{5}{12}$ and $\sin B = \frac{15}{17}$ find:

6. tan(A+B) if A is acute and B obtuse 7. cos3A if A is in the third quadrant

8. sin½A if A is acute 9. sin½A if A is in the third quadrant

10. sec(60°– ½B) if B is acute

INVERSE FUNCTIONS

Example 8 Evaluate $\tan^{-1}\frac{1}{2} + \tan^{-1}\frac{1}{3}$.
(That is, the angle whose tangent is $\frac{1}{2}$ plus the angle whose tangent is $\frac{1}{3}$.)

We have:

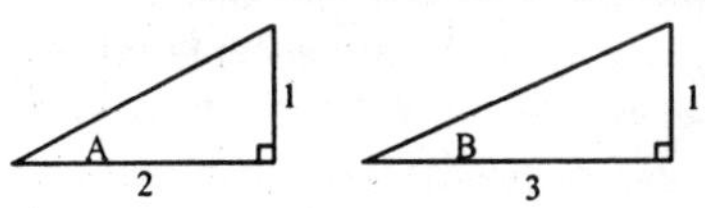

where $A = \tan^{-1}\frac{1}{2}$
and $B = \tan^{-1}\frac{1}{3}$

$\therefore$

A	2	1	-	
B	3	1	-	+
A+B	5	5	-	

$= 1, 1, \sqrt{2}$

So $\tan^{-1}\frac{1}{2} + \tan^{-1}\frac{1}{3} = \frac{\pi}{4}$ radians.

Example 9 Show that $2\tan^{-1}2 + \tan^{-1}3 = \pi + \tan^{-1}\frac{1}{3}$

Let the angles be A, B and C respectively.
So we have to prove that $2A + B = \pi + C$.

Then

A	1	2	-	
2A	−3	4	-	
B	1	3	-	+
2A+B	−15	−5	-	
=	−3	−1	-	

But

π	−1	0	1	
C	3	1	-	+
π +C	−3	−1	-	

$\therefore\ 2\tan^{-1}2 + \tan^{-1}3 = \pi + \tan^{-1}\frac{1}{3}$

It is worth noting here that the abbreviations sin, cos, tan, cosec, sec, cot are avoided by this triple method. Students find this use of these "part-words" confusing: the triple notation is much more straightforward and clear.
Conventional trigonometry also relies on a mass of complex formulae which the student has to remember or look up. These are not needed in the triple method.

THE GENERAL TRIPLE

Since dividing a triple through by any positive number does not change the angle

$$A)\ x \quad y \quad 1$$

can be considered a general triple capable of representing any angle.

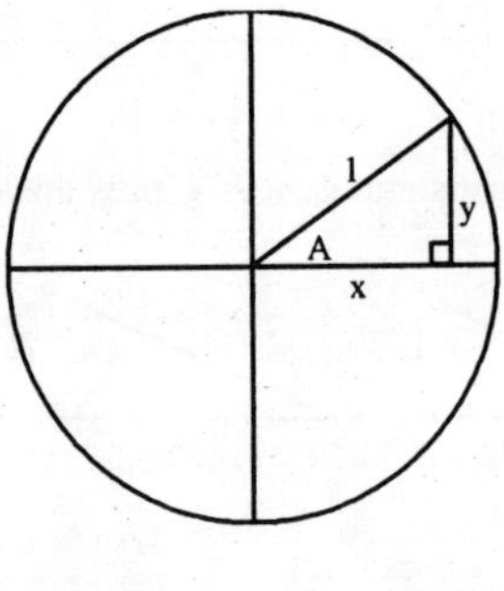

Then $\cos A = x$ $\sin A = y$ $\tan A = \frac{y}{x}$ $\cot A = \frac{x}{y}$ $\sec A = \frac{1}{x}$ $\text{cosec} A = \frac{1}{y}$

These can be neatly represented on a single diagram:

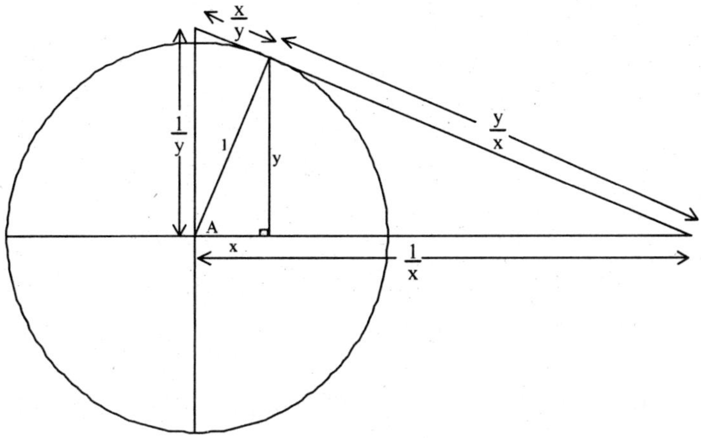

Or, alternatively:

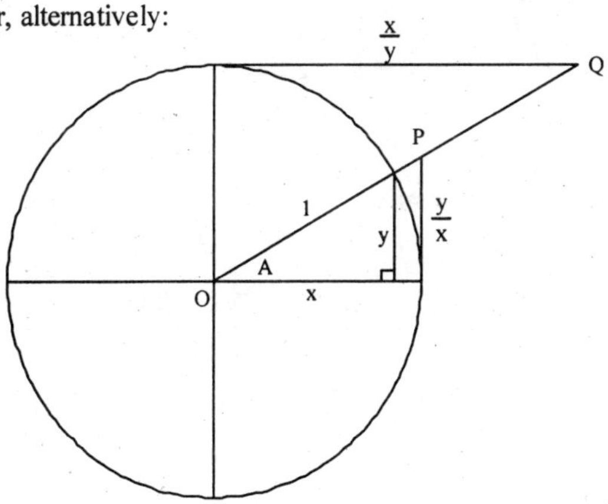

where $OP = \frac{1}{x}$ and $OQ = \frac{1}{y}$.

So $x^2 + y^2 = 1$.
And the triple for the general double angle is also frequently in use:

A	x	y	r
2A	$(x^2 - y^2)$,	$(2xy)$,	(r^2)

<u>Example 10</u> Prove that tanA + cotA ≡ secAcosecA.

That is: prove $\frac{y}{x} + \frac{x}{y} = \frac{1}{x} \times \frac{1}{y}$

But $\frac{y}{x} + \frac{x}{y} = \frac{y^2+x^2}{xy} = \frac{1}{xy}$ as $x^2+y^2=1$

$\therefore \frac{y}{x} + \frac{x}{y} = \frac{1}{x} \times \frac{1}{y}$ Q.E.D.

Example 11 Prove that $\sin^2 A \operatorname{cosec}(90° - A) \equiv \cot^2(90° - A)\cos A$

We need a triple for $90° - A$:

90°	0	1	1	
A	x	y	1	−
90°−A	y	x	1	

So we have to show that $y^2 \times \frac{1}{x} = \frac{y^2}{x^2} \times x$.

Which it clearly is, so Q.E.D.

Example 12 Prove $\frac{\sin 2A}{1+\cos 2A} \equiv \tan A$

Using the triples

A	x	y	r
2A	$(x^2 - y^2)$,	$(2xy)$,	(r^2)

we need to prove that $\frac{2xy}{1+x^2-y^2} = \frac{y}{x}$

But $1 - y^2 = x^2 \ \therefore \ \frac{2xy}{1+x^2-y^2} = \frac{2xy}{2x^2} = \frac{y}{x}$ Q.E.D.

Example 13 Express $\sqrt{\frac{1-\sin 2A}{1+\sin 2A}}$ in terms of tanA.

$$\sqrt{\frac{1-\sin 2A}{1+\sin 2A}} = \sqrt{\frac{1-2xy}{1+2xy}} = \sqrt{\frac{x^2+y^2-2xy}{x^2+y^2+2xy}} = \frac{x-y}{x+y} = \frac{1-\frac{y}{x}}{1+\frac{y}{x}} = \frac{1-\tan A}{1+\tan A}$$

Example 14 Simplify $\tan^{-1} x + \tan^{-1}\left(\frac{1-x}{1+x}\right)$.

We have

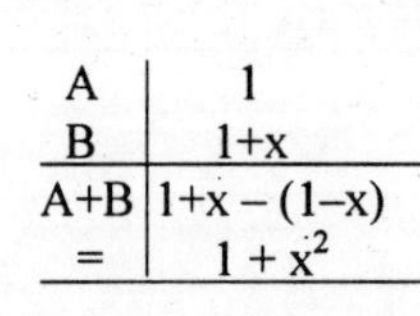

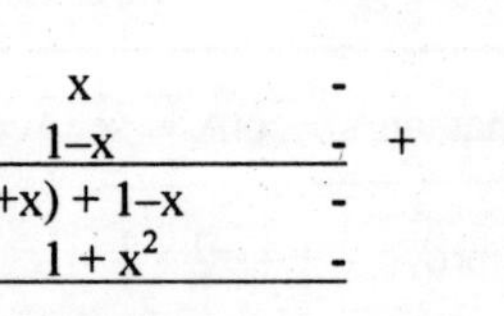

We are given the addition of two angles so we can add their triples:

A	1	x	-	
B	1+x	1−x	-	+
A+B	1+x − (1−x)	x(1+x) + 1−x	-	
=	$1 + x^2$	$1 + x^2$	-	

Since the first and second elements of the result are equal this triple must be 1, 1, $\sqrt{2}$ (we could divide the triple through by $1+x^2$). The angle in this triple must therefore be 45° or $\frac{\pi}{4}$ radians.

So $\tan^{-1} x + \tan^{-1}\left(\frac{1-x}{1+x}\right) = \frac{\pi}{4}$.

<u>Example 15</u> Show $\sin(2\tan^{-1}x) = \frac{2x}{x^2+1}$.

Let $\tan^{-1}x = A$:

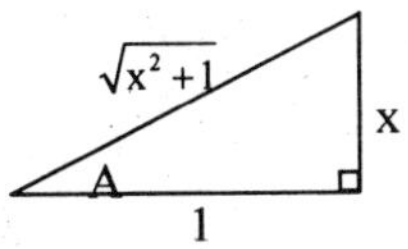

Then

A	1	x	$\sqrt{x^2+1}$
2A	-	2x	x^2+1

then $\sin(2\tan^{-1}x) = \sin 2A = \frac{2x}{x^2+1}$

<u>Example 16</u> Prove that $\tan A + \tan B = \frac{\sin(A+B)}{\cos A \cos B}$

Let A) x y 1
B) X Y 1

$$\begin{array}{ccc} x & y & 1 \\ X & Y & 1 \\ \hline xX-yY & yX+xY & 1 \end{array} +$$

Then $\frac{\sin(A+B)}{\cos A\cos B} = \frac{yX+xY}{xX} = \frac{y}{x} + \frac{Y}{X} = \tan A + \tan B$.

SOLUTION OF TRIGONOMETRICAL EQUATIONS

<u>Example 17</u> Express in triple form: a) $\sin A = \frac{1}{2}$ b) $\cos B = \frac{1}{2}$ c) $\tan C = \sqrt{3}$

a) If $\sin A = \frac{1}{2}$ then A) - , 1, 2, and so A) $\pm\sqrt{3}$, 1, 2 are the triples.

b) Similarly, if $\cos B = \frac{1}{2}$ then B) 1, $\pm\sqrt{3}$, 2.

c) $\tan C = \sqrt{3} = \frac{\sqrt{3}}{1}$ so C) ± 1, $\pm\sqrt{3}$, 2 (two triples)

<u>Example 18</u> Solve $\cos 2A - \sin 2A = 1$.

With the usual substitutions we have $x^2 - y^2 - 2xy = 1$.

$\therefore -y^2 - 2xy = y^2$ $\therefore -2xy = 2y^2$ $\therefore y = 0$ or $\frac{y}{x} = -1$.

y=0 gives	± 1	0	1	
and $\frac{y}{x} = -1$ gives	-1	1	$\sqrt{2}$	
and	1	-1	$\sqrt{2}$	four solutions in all.

Example 19 Solve $3\cot^2 A + 5 = 7\operatorname{cosec}A$.

That is, $\frac{3x^2}{y^2} + 5 = \frac{7}{y}$

$\therefore\ 3x^2 + 5y^2 = 7y \quad \therefore\ 3 - 3y^2 + 5y^2 = 7y \quad \therefore\ 2y^2 - 7y + 3 = 0$

$\therefore\ (2y - 1)(y - 3) = 0$

$\therefore\ y = \frac{1}{2}$ and so A) $\pm\sqrt{3}$, 1, 2.

Exercise B

1. Evaluate $\tan^{-1}\frac{1}{2} + \tan^{-1}\frac{1}{5} + \tan^{-1}\frac{1}{8}$

Prove 2. $2\tan^{-1}\frac{1}{2} = \tan^{-1}\frac{4}{3}$

3. $(\operatorname{cosec}^2 A - 1)\sin^2 A = \cos^2 A$

4. $\tan A(1 + \cos 2A) = \sin 2A$

5. $(1 + \sin A + \cos A)^2 = 2(1 + \sin A)(1 + \cos A)$

6. $\sin 2A = \frac{2\tan A}{1+\tan^2 A}$

7. $\sin(120° - A) = \sin(60° + A)$

8. $\cot(A+B) = \frac{\cot A \cot B - 1}{\cot A + \cot B}$

9. $\frac{\sin A}{1+\cos A} = \tan\frac{1}{2}A$

10. $\sec 2A + \tan 2A = \tan(45° + A)$

11. $3\tan^{-1} 2 - \pi = \tan^{-1}\frac{2}{11}$

12. $4\tan^{-1}\frac{1}{5} - \tan^{-1}\frac{1}{239} = \frac{\pi}{4}$ (this is Machin's formula, useful for evaluating π)

13. $4(\cos^4 A + \sin^4 A) = 3 + \cos 4A$

14. $\frac{\sin 2A - \cos 2A + 1}{\sin 2A + \cos 2A + 1} = \tan A$

15. If $\tan(A+B) = \frac{1}{2}$ and $\tan(A-B) = \frac{1}{3}$ find tan2A and A.

continued

Solve the following equations giving the answers as triples:

16. $\sin A(\sin A + \cos A) = 1$

17. $1 - \sin A = 2\cos^2 A$

18. $\tan 2A = 4\tan A$.

19. $3\cot 2A + 7\tan A = 5\operatorname{cosec} 2A$.

20. $\tan^{-1}\frac{1}{2} - \tan^{-1}\frac{1}{3} = \sin^{-1} x$

There are various other useful triple methods for solving trigonometrical equations as the following three sections show.

FURTHER TRIGONOMETRICAL EQUATIONS: A

Example 20 Solve $10\cos A + 5\sin A = 11$.

This converts to $10x + 5y = 11$.
From the coefficients on the left of this equation we construct the triple: B) $10, \; 5, \; -$.

Now consider the following triple subtraction:

$$\begin{array}{r|ccc}
A & x & y & 1 \\
B & 10 & 5 & \sqrt{125} \quad - \\
\hline
A-B & 11 & & \\
\hline
\end{array}$$

The two vertical products formed when beginning this subtraction sum give $10x + 5y$ and since we know that this is 11 we can put this down in the first column of the answer.

Also we know the last element of the answer: $\sqrt{125}$. We can therefore fill in the middle element, which must be 2.
So A–B) $11, \; 2, \; \sqrt{125}$.
We can now add the second and third rows and since we are adding the angles B and A–B the result must be a triple for A, which is what we require:

$$\begin{array}{r|ccc}
A & x & y & 1 \\
B & 10 & 5 & \sqrt{125} \quad - \\
\hline
A-B & 11 & 2 & \sqrt{125} \quad + \\
\hline
A & 100 & 75 & 125 = 4,3,5
\end{array}$$

which is an answer.

Now in evaluating the middle element of the triple for A–B the result could have been –2 as well as 2. But adding the triples with 2 changed to –2 will give the same result as subtracting the original triples:

A	x	y	1	
B	10	5	$\sqrt{125}$	–
A – B	11	2	$\sqrt{125}$	–
A	120	35	125 = 24,7,25	which is the other solution.

Alternatively of course 2 and –2 can be used as the middle element of A–B and only triple addition need then be used.
There are variations of this method but the one shown above is the most efficient.

Example 21 Solve 4cosA – 7sinA = 8.

This becomes 4x – 7y = 8.
The coefficients 4, –7 are inserted as the first elements of B, 8 is then the first entry for A–B:

A :	x	y	1		
B :	4	–7	$\sqrt{65}$	–	
A–B :	8	±1	$\sqrt{65}$	–	
A :	39	–52	65	=	3, –4, 5
A :	25	–60	65	=	5, –12, 13

The method is just the same. We add and then subtract in the third row to get the two answers.

It is worth noting that the third element of the triple for B gives the maximum and minimum values of the expression on the left-hand side of the equation: $\pm\sqrt{65}$ in the example above.

This example shows the whole working and is far easier than the conventional method for solving this type of equation. See Chapter 13 for a comparison of this with the traditional method.

An alternative solution involving triple code numbers will be shown later.

FURTHER TRIGONOMETRICAL EQUATIONS: B

Example 22 Solve sin2A + sin4A = 0

Let 2A) x, y, 1.
The above equation says that the middle element of the triple for 4A is minus that of the triple for 2A. So we can write:

2A) x y 1
4A) - –y 1

We can then find the first element of the triple for 4A from the other two elements to get:

$$\begin{array}{llll} 2A) & x & y & 1 \\ 4A) & \pm x & -y & 1 \end{array}$$

We now seek to eliminate the letters x and y *By Addition and By Subtraction* of these two triples. Taking the first element of 4A as +x first, and adding:

$$\begin{array}{c|ccc} 2A & x & y & 1 \\ 4A & x & -y & 1 \; + \\ \hline 6A & 1 & 0 & 1 \end{array} \qquad \text{as } x^2 + y^2 = 1$$

Since 1,0,1 means 360° it follows that <u>A = 60n°</u>.

(Adding the above triples with –x as the first element for 4A does not give any new solutions. Similarly subtracting the triples with x as the first element for 4A gives no new solutions.)

If we now subtract the triples above:

$$\begin{array}{c|ccc} 4A & -x & -y & 1 \\ 2A & x & y & 1 \; - \\ \hline 2A & -1 & 0 & 1 \end{array} \qquad \text{as } -x^2 - y^2 = -1$$

–1,0,1 represents 180° + 360n° so <u>A = 90° + 180n°</u>.
These two answers give all the solutions to the given equation.

A = 60n° could be given in triple form as $\pm 1 \quad \pm\sqrt{3} \quad 2$
and $\pm 1 \quad 0 \quad 1$ (which is not as compact as 60n).
A = 90° + 180n° would be written as $0 \quad \pm 1 \quad 1$ (which is better than 90°+180n°).

This same equation is solved by a different method in Example 25.

<u>Example 23</u> Solve cos2A = sin3A

Let 2A) x y 1. Then the above equation gives us 3A) - , x, 1:

$$\begin{array}{c|ccc} 3A & \pm y & x & 1 \\ 2A & x & y & 1 \; + \\ \hline 5A & 0 & 1 & 1 \end{array}$$

Taking the first element of 3A to be +y we have to add the triples to eliminate the letters. With –y instead we have to subtract them; but in both cases we get 5A)0 1 1.

Since 0, 1, 1 is the triple for 90° we have (working in radians):

$$5A = \tfrac{\pi}{2} + 2\pi n \qquad \text{so that} \qquad A = \tfrac{\pi}{10} + \tfrac{2\pi n}{5}$$

The reader is invited to solve this equation without using triples.

FURTHER TRIGONOMETRICAL EQUATIONS: C

If we add and subtract the triples 4,3,5 and 12,5,13 we get the following:

A	4	3	5	
B	12	5	13	±
A+B	33	56	65	
A–B	63	16	65	

From this we may note the following relationships between the various elements:

$$2(4 \times 12) = 33 + 63 \qquad (1)$$
$$2(3 \times 5) = 63 - 33 \qquad (2)$$
$$2(3 \times 12) = 56 + 16 \qquad (3)$$
$$2(4 \times 5) = 56 - 16 \qquad (4)$$

These relationships are generally true and we can make use of them to solve trigonometrical equations which contain multiple angles.

Example 24 Solve cosA = cos2A + cos4A.

First consider the table:

3A	a	b	1
A	c	d	1
4A	e	f	1
2A	g	h	1

in which we have added the first two rows to get the 3rd row and subtracted the first two rows to get the 4th row.

We know from the results above that there are certain relationships between the elements a,b,c,d,e,f,g,h.
We know for example e + g = 2ac (result (1) above).
But from the original equation e + g = c.
So we conclude 2ac = c which means that c=0 or a=½.

c=0 means that A) 0, ± 1, 1 and so A = 90°+ 180n°.

a=½ means that 3A) ½, - , 1 = 1, - , 2 and so 3A = ± 60° + 360n°
and therefore A = ±20°+120n°.

Example 25 Solve sin2A + sin4A = 0.

Since we have got a sum here we must arrange to have the triples for 2A and 4A on the last two lines if we are going to make use of the triple element relationships noted above.

3A	a	b	1
A	c	d	1
4A	e	f	1
2A	g	h	1

The given equation tells us that f + h = 0, but we also know that f + h = 2bc (result (3)).
So 2bc = 0 and b=0 or c=0.
If c=0, A) 0, ± 1, 1 or <u>A = 90° + 180n°</u>.
If b=0, 3A) ± 1, 0, 1 so that 3A=180n° and <u>A = 60n°</u>.

<u>Example 26</u> Express 2sin15°sin45° in terms of surds.

45°	-	a	1
15°	-	b	1
60°	d	-	1
30°	c	-	1

From result (2) above we know that 2ab = c – d, and 2ab is the required quantity, 2sin15sin45°. But c and d are known (see previous chapter): these are $\frac{\sqrt{3}}{2}$ and $\frac{1}{2}$
So 2sin15°sin45° = $\frac{\sqrt{3}}{2} - \frac{1}{2}$

A further triple method for solving trigonometrical equations (using the triple code numbers) will be found in Chapter 9.

<u>Exercise C</u> Solve the following equations:

1. 10 cosA – 5sinA = 11   2. 2cosA + 3sinA = 3   3. 4cosA + 7sinA = 8

4. sin5A + sin3A = 0 (solve 2 ways, as in sections B and C above)

5. cos2A + sin4A = 0   6. cos3A + cosA = cos2A

7. Evaluate 2cos15° sin75°

4 TRANSFORMATIONS IN A PLANE

In this chapter we deal mainly with rotations and reflections in two dimensions.
We saw in Chapter 1 that triples can be multiplied or divided by any positive number without changing the angle in the triple and so far we have always tried to express a triple in its simplest form. Now, however, because we are considering actual lengths, we must be ready to multiply or divide the triples as necessary.

Before proceeding let us look again at the subtraction of triples. This was described in Chapter 2 but there is another geometrical way of subtracting two triples:

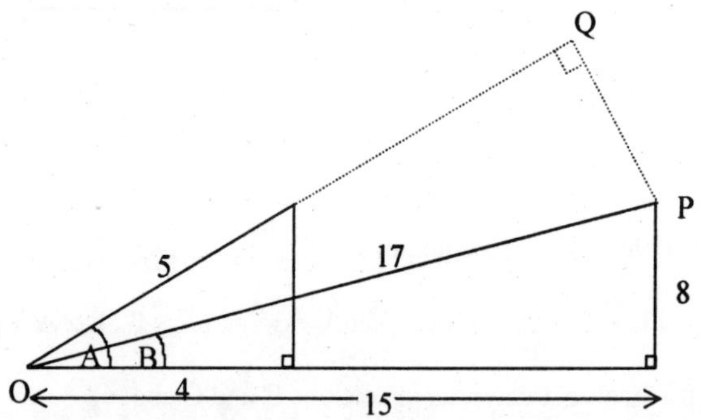

Here we have A) 4, 3, 5 and B) 15, 8, 17 and both triangles have their base on the same line.
The angle A–B is therefore angle POQ.
The calculation is the same as in Chapter 2:

A	4	3	5
B	15	8	17 –
A – B	84	13	85

But if we want the actual dimensions of triangle OPQ we must divide 84,13,85 through by 5 because the hypotenuse must be 17: this gives $\frac{84}{5}$, $\frac{13}{5}$, 17.

The first example using subtraction like this is Example 18 in this chapter.

TRANSPOSITION OF THE ORIGIN

It is often convenient to transpose the origin to another position.
The coordinates of all points therefore change.
For example, the point (7,5) has coordinates (5,2) if the origin is transposed to the point (2,3):

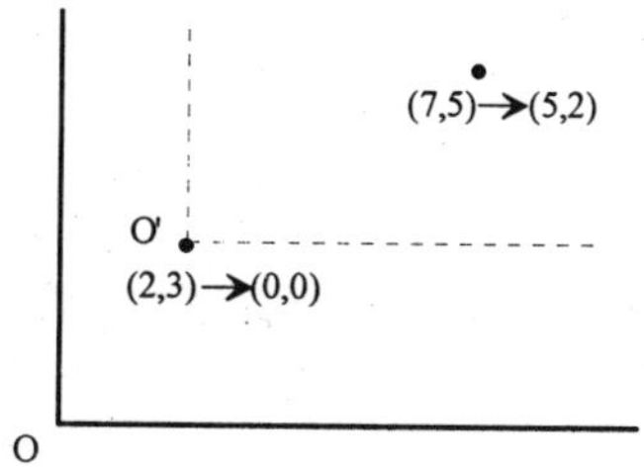

In general if the origin is transposed to a point, the coordinates of that point are subtracted from those of all other points.

Example 1 Find the coordinates of the points P(4,7), Q(–3,4), R(0,–5) when the origin is transposed to (3, –1).

We subtract 3 and –1 from each pair of coordinates: P'(1,8), Q'(–6,5), R'(–3,–4).

ROTATIONS

We will assume all rotations to be anticlockwise unless otherwise stated.

Example 2 Rotate P(7,3) through an angle of 90° using the origin as centre of rotation.

We are effectively adding two triangles, one a 7, 3, $\sqrt{58}$ triangle and the other 0,1,1 (triple for 90°)

7	3	$\sqrt{58}$	
0	1	1	+
–3	7	$\sqrt{58}$	

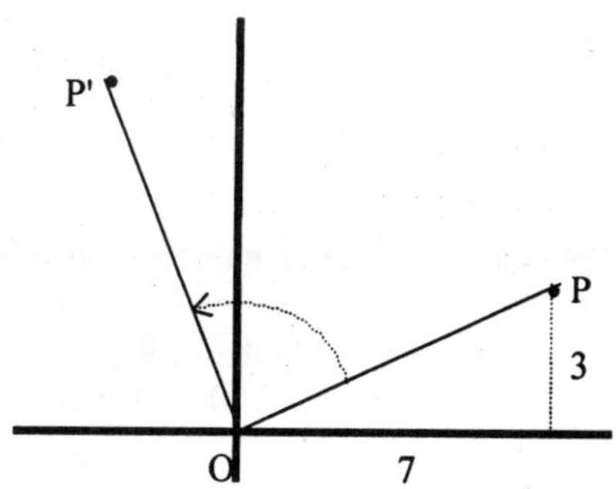

So the position of the rotated point, P', is (–3,7).

Note that since OP=OP' the third column is superfluous.

Example 3 Rotate P(–2,5) 90° clockwise about the point (1,7).

First we transpose the origin to (1,7) and so consider the rotation of (–3,–2) 90° clockwise about (0,0). A clockwise rotation will involve **subtracting** the triples:

$$\begin{array}{ccc} -3 & -2 & - \\ \underline{0} & \underline{1} & \underline{-} \quad - \\ \underline{-2} & \underline{3} & \underline{-} \end{array}$$

So the rotated point is (–2,3) relative to (1,7) as origin.
We return to the original origin by **adding** (1,7) back on to get P'(–1,10).

Example 4 Rotate the point P(5,2) through an angle A) 4,3,5 about the origin.

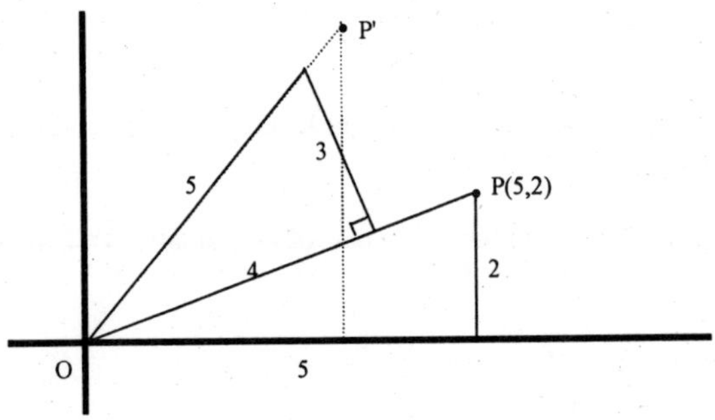

The point (5,2) gives us a right-angled triangle: 5, 2, $\sqrt{29}$.
As the diagram shows, rotating through A) 4,3,5 is equivalent to adding these triples.

$$\begin{array}{ccc} 5 & 2 & \sqrt{29} \\ \underline{4} & \underline{3} & \underline{5} \quad + \\ \underline{14} & \underline{23} & \underline{5\sqrt{29}} \end{array}$$

But we want the hypotenuse of the resulting triangle to be equal to OP, which is $\sqrt{29}$.
We therefore divide through by 5 which gives P' $(\frac{14}{5}, \frac{23}{5})$ or P'(2.8, 4.6).

Note that the third column in the calculation above is unnecessary as we consistently divide through by the third element of the rotation triple.

If P is a "point triple" and R is a "rotation triple" then the above method can be described by the equation P' = P + R where the addition is triple addition.
Similarly P' = P + R_1 + R_2 means we rotate P by R_1 and then by R_2.

Also $P' = P + (R_1 + R_2)$ shows that rotation triples can be added to get a triple which combines R_1 and R_2.

Repeated rotations can be used to create an appearance of motion when the images are quickly shown and removed. The diagrams on this page are composed of rotating lines. They were drawn with the aid of a BBC microcomputer and the programs used the triple method of rotation described in this chapter.
The motif at the beginning of each chapter consists of a 5,12,13 triple rotated by the triple 99,20,101.

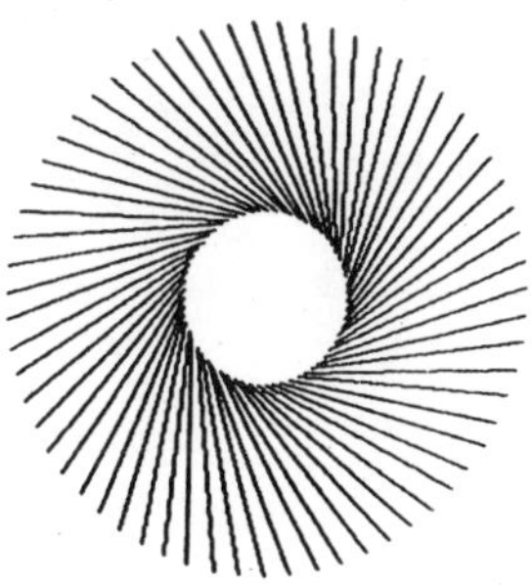

Example 5 Rotate P(3,5) through A) –4,3,5 about the origin.

Here we are rotating by an obtuse angle but the method is the same:

$$\begin{array}{ccc} 3 & 5 & - \\ \underline{-4} & \underline{3} & \underline{5} \quad - \\ \underline{-27} & \underline{-11} & \underline{-} \end{array}$$

∴ P'(–5.4,–2.2).

Example 6 Rotate P(4,–5) clockwise through A) 4,3,5 about O.

We subtract the triples for clockwise rotation:

$$\begin{array}{ccc} 4 & -5 & - \\ \underline{4} & \underline{3} & \underline{-} \quad - \\ \underline{1} & \underline{-32} & \underline{-} \end{array}$$

∴ P'(0.2, –6.4).

Example 7 Rotate P(4,7) through A) 3,4,5 about (2,3).

First we transpose the origin to (2,3) and rotate (2,4) by A) 3,4,5:

$$\begin{array}{ccc} 2 & 4 & - \\ \underline{3} & \underline{4} & \underline{5} \quad + \\ \underline{-10} & \underline{20} & \underline{-} \end{array}$$

So the rotated point (relative to (2,3) as origin) is (–2,4).
And returning to the original origin by adding (2,3) back on we get P'(0,7).

Example 8 Rotate P(8,–5) 135° about the point (7,7).

Transposing the origin we rotate (1,–12) by the triple $-1, 1, \sqrt{2}$.

$$\begin{array}{ccc} 1 & -12 & - \\ \underline{-1} & \underline{1} & \underline{\sqrt{2}} \quad + \\ \underline{11} & \underline{13} & \underline{-} \end{array}$$

∴ The transposed rotated point is $\left(\frac{11}{\sqrt{2}}, \frac{13}{\sqrt{2}}\right)$.

And the rotated point is $P'\left(\frac{11}{\sqrt{2}} + 7, \frac{13}{\sqrt{2}} + 7\right)$.

SPIRALS

Suppose we start with the point (1,0) and rotate it through an angle of 45° about the origin by adding the triple $1,1,\sqrt{2}$:

$$\begin{array}{ccc} 1 & 0 & 1 \\ 1 & 1 & \sqrt{2} \; + \\ \hline 1 & 1 & \sqrt{2} \\ \hline \end{array}$$

then the point (1,1) obtained is distant $\sqrt{2}$ from the origin.

If (1,1) is then rotated in a similar manner:

$$\begin{array}{ccc} 1 & 1 & \sqrt{2} \\ 1 & 1 & \sqrt{2} \; + \\ \hline 0 & 2 & 2 \\ \hline \end{array}$$

we get the point (0,2) which is 2 units from the origin.

Continuing to add $1,1,\sqrt{2}$ and leaving aside the third element in each triple we get:

$$\begin{array}{cc} 0 & 2 \\ 1 & 1\,+ \\ \hline -2 & 2 \\ \hline \end{array} \qquad \begin{array}{cc} -2 & 2 \\ 1 & 1\,+ \\ \hline -4 & 0 \\ \hline \end{array} \qquad \begin{array}{cc} -4 & 0 \\ 1 & 1\,+ \\ \hline -4 & -4 \\ \hline \end{array} \qquad \begin{array}{cc} -4 & -4 \\ 1 & 1\,+ \\ \hline 0 & -8 \\ \hline \end{array} \qquad \begin{array}{cc} 0 & -8 \\ 1 & 1\,+ \\ \hline 8 & -8 \\ \hline \end{array}$$

Plotting all these points
Gives the equiangular spiral:

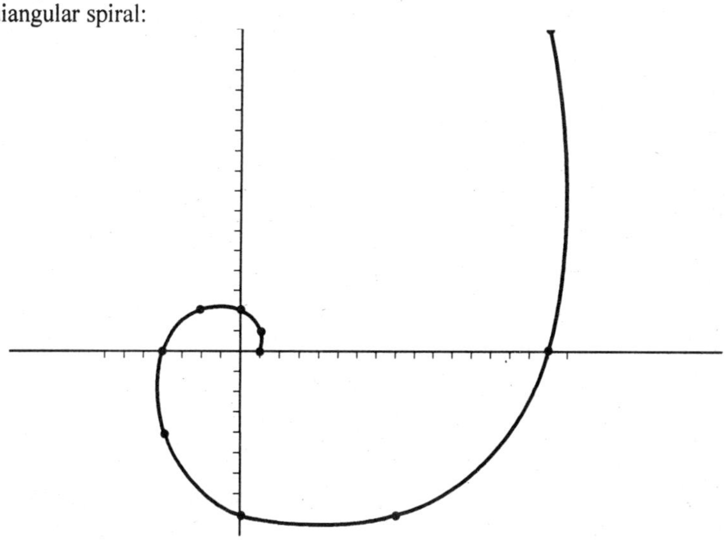

INTEGRATION OF COSX ETC.

Since successive integrals of the function cosx effectively advance the curve through 90° each time, the integrals and derivatives of cosx, sinx, –cosx, –sinx can be shown as follows:

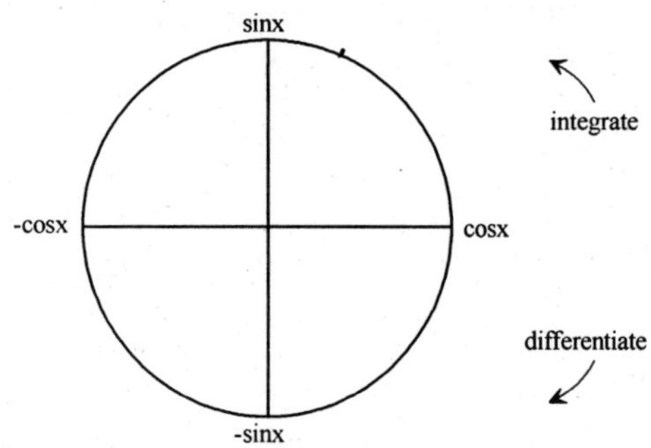

Example 9 $\int \sin x dx = -\cos x + c$. We start at sinx and take one step anticlockwise.

Example 10 $\frac{d^2(\cos x)}{dx^2} = -\cos x$. Similarly we take two clockwise steps.

Exercise A Rotate:

1. (3,5) through an angle of 90° about the origin.
2. (–7,2) 180° about the origin.
3. (11,4) 90° about the point (3,5).
4. (1,–2) 270° about the point (5,5).
5. (2,1) by the angle 7,24,25 about the origin.
6. (–2,3) by 3,4,5 about the origin.
7. (0,–3) by 4,–3,5 clockwise about the origin.
8. (1,–3) by 4,3,5 about the point (–2,3).
9. (–3,1) by 3,–4,5 about (1,2).
10. (–4,6) by 30° about (0,0).
11. Through what angle (triple) must the point (10,3) be rotated about the origin so that it is in the direction OP where P is the point (5,4)?

ROTATION OF LINES AND CURVES

Example 11 Find the equation of the line $y = 2x + 1$ when it has been rotated through the angle 4,3,5 about the origin.

Let (x,y) and (X,Y) represent corresponding points on the given line and on the rotated line respectively.
Since both points are the same distance from the origin they can be represented by the triples x,y,z and X,Y,z.

Then
$$\begin{array}{ccc} X & Y & z \\ 4 & 3 & 5 \ - \\ \hline x & y & z \\ \hline \end{array}$$
But
$$\begin{array}{ccc} X & Y & z \\ 4 & 3 & 5 \ - \\ \hline \frac{4X+3Y}{5} & \frac{4Y-3X}{5} & z \\ \hline \end{array}$$

$\therefore\ x = \frac{4X+3Y}{5}$ and $y = \frac{4Y-3X}{5}$.

So for $y = 2x + 1$ we may write: $\frac{4Y-3X}{5} = 2\times\frac{4X+3Y}{5} + 1$ as the equation of the rotated line.

This simplifies to $4Y - 3X = 8X + 6Y + 5$ and to $2Y + 11X + 5 = 0$.

which may be written as $2y + 11x + 5 = 0$.

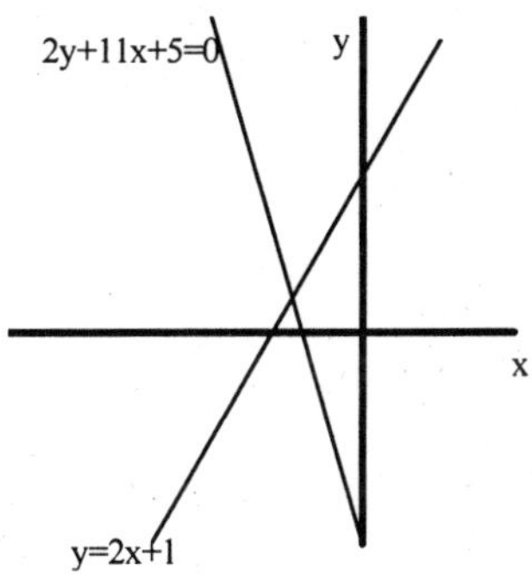

Example 12 Rotate $2x + 3y = 4$ through the angle 3,4,5.

$$\begin{array}{ccc} X & Y & z \\ 3 & 4 & 5 \ - \\ \hline \frac{3X+4Y}{5} & \frac{3Y-4X}{5} & z \\ \hline \end{array}$$
so we substitute $\frac{3X+4Y}{5}$ for x and $\frac{3Y-4X}{5}$ for y:

$\therefore\ 2\times\frac{3X+4Y}{5} + 3\times\frac{3Y-4X}{5} = 4$ which leads to $6x - 17y + 20 = 0$.

Example 13 Find the equation of the curve $y = x^2$ after a clockwise rotation of 90° about the origin.

The method is just the same:

$$\begin{array}{ccc} X & Y & z \\ 0 & 1 & 1 \quad + \\ \hline -Y & X & z \end{array}$$

So we replace x with –Y and y with X to get $X = (-Y)^2$. $\therefore$ $\underline{x = y^2}$.

Example 14 Find the equation of $y = x^2$ when rotated by 4,3,5.

The substitutions for a rotation by 4,3,5 were found in Example 11, so we get:

$$\frac{4Y-3X}{5} = \left(\frac{4X+3Y}{5}\right)^2$$

$\therefore$ $20Y - 15X = 16X^2 + 24XY + 9Y^2$

$\therefore$ $\underline{16x^2 + 9y^2 + 24xy + 15x - 20y = 0}$ is the required equation.

Example 15 Find the equation for $y = \frac{1}{x}$ when rotated 45° about the origin.

$$\begin{array}{ccc} X & Y & z \\ 1 & 1 & \sqrt{2} \quad - \\ \hline \frac{X+Y}{\sqrt{2}} & \frac{Y-X}{\sqrt{2}} & z \end{array} \qquad \therefore \quad \frac{Y-X}{\sqrt{2}} = \frac{\sqrt{2}}{X+Y}$$

So the equation is $\underline{y^2 - x^2 = 2}$.

When rotating straight lines the equation of the rotated line can be found more easily as follows.

For Example 11:

$$\begin{array}{ccccc} 1y & = & 2x & + & 1 \\ 4 & & 3 & & 5 \quad + \\ \hline \multicolumn{5}{c}{(1\times4 - 2\times3)y = (2\times4 + 1\times3)x + 1\times5} \end{array}$$

$\therefore$ $\underline{-2y = 11x + 5}$.

For Example 12:

$$\begin{array}{ccccc} 2x & + & 3y & = & 4 \\ 3 & & 4 & & 5 \quad + \\ \hline \multicolumn{5}{c}{(2\times3 - 3\times4)x + (3\times3 + 2\times4)y = 4\times5} \end{array}$$

$\therefore$ $\underline{-6x + 17y = 20}$.

We see that the same *Vertically and Crosswise* method gives the correct answer and that it does not matter whether the equation of the line is of the form of Example 11 or 12.
This result can easily be proved with a little algebra. Note, however, that for these anticlockwise rotations the rotation triple is added, not subtracted as in the other examples.

Later we will see how to rotate curves in 3-dimensional space.

REFLECTIONS

There are four main types of reflections: reflection in:

1. one of the coordinate axes,
2. a line parallel to one of the axes,
3. an inclined line through the origin,
4. an inclined line not through the origin.

<u>Example 16</u> Reflect P(–10,7) in the x-axis.

The x-coordinate remains unchanged and the y-coordinate has its sign transposed.
That is: 7 becomes –7.

Therefore P'(–10,–7).

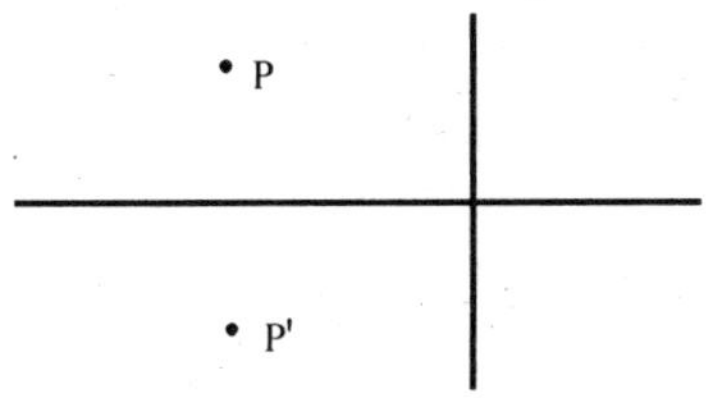

Similarly, for reflections in the y-axis the x-coordinate is transposed.

<u>Example 17</u> Reflect P(9,–5) in the line x=4.

We transpose the origin to (4,0) and reflect (5,–5) in x=0 to get (–5,–5).

Then add x=4 back on to get P'(–1,–5).

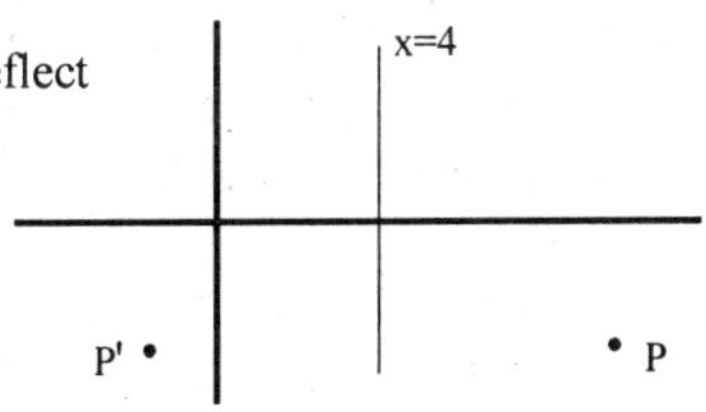

<u>Example 18</u> Reflect P(5,2) in the line 4y=3x.

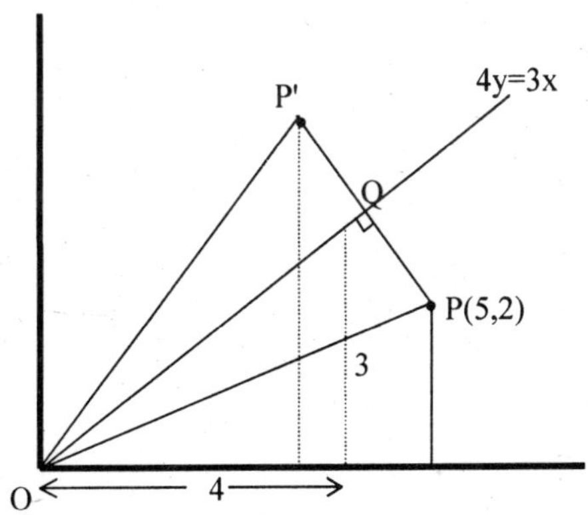

We require the coordinates of P' where $P\hat{O}Q = P'\hat{O}Q$ and OP=OP'.
4,3,5 is a triple representing the line 4y = 3x.
If we now subtract the triple for the point from the triple for the line, we will get a triple for the remainder, i.e. for angle POQ.

We can then add this on to the line triple in order to perform the reflection.

line triple:	4	3	5	
P:	5	2	$\sqrt{29}$	–
	26	7	$5\sqrt{29}$	
	4	3	5	+
	83	106	$25\sqrt{29}$	

Since OP = OP' = $\sqrt{29}$ we divide through by 25.

$\therefore$ P'$\left(\frac{83}{25}, \frac{106}{25}\right)$.

As with rotations the third column can be dispensed with as we always divide by the square of the third element of the line triple.

Note also that when the equation of the line is in the form ay = bx then a and b are the first two elements of the triple for the line.

Example 19 Reflect P(5,5) in the line y = 2x.

Similarly, line triple:	1	2	$\sqrt{5}$	
P:	5	5	-	–
	15	5	-	
	1	2	-	+
	5	35	-	

We divide (5,35) by the square of the third element of the line triple, i.e. by 5.

$\therefore$ P'(1,7).

It may be seen from these last two examples that we subtract the point triple from twice the line triple.

E.g.	line triple:	1	2	$\sqrt{5}$	
	2 × line triple:	–3	4	5	
	P:	5	5	-	–
		5	35	-	$\therefore$ P'(1,7) as before.

Using the notation introduced in the rotations section we could write this as: P' = 2L – P.

Example 20 Reflect P(4,9) in the line y = 2x + 3.

Since this line does not pass through the origin we transpose the origin to the point (0,3) which is on the line and consider the reflection of (4,6) in y = 2x:

line:	1	2	$\sqrt{5}$	
2 × line:	–3	4	-	
point:	4	6	-	–
	12	34	-	

Then dividing by 5 the transposed reflected point is (2.4,6.8).
Adding back 3 to the y-coordinate gives: P'(2.4,9.8).

Example 21 Reflect P(–2,3) in the line 3y = 4x + 2.

We move the origin to a suitable position: (1,2) is on the line.
Then reflect (–3,1) in 3y = 4x:

2 × line:	–7	24	25	
point:	–3	1	-	–
	45	–65	-	

∴ the transposed reflection is at (1.8,–2.6) and P'(2.8,–0.6).

Exercise B Reflect:

1. P(4,7) in the y-axis
2. P(–3,–8) in the x-axis
3. P(8,2) in the line x=5.
4. P(3,4) in y = 2x
5. P(12,5) in 4y = 3x
6. P(–5,10) in 3y = x
7. P(4,–5) in y = x
8. P(2,5) in y = 2x
9. P(1,3) in y = 2x + 5
10. P(3,8) in 4y = –3x + 1

Transformations can be compounded so that, for example, a single triple can be found for a rotation followed by a reflection.
If we want to reflect a point P in a line L and then rotate the result by an amount R, where P, L, R are the point, line and rotation triples respectively, then P' = (2L – P) + R = (2L + R) – P.
That is, (2L + R) effectively combines the two transformations.
For a rotation followed by a reflection, P' = 2L – (P + R) = (2L – R) – P.

Standard matrix methods for rotations and reflections require obtaining or remembering the formulae for rotating and reflecting as well as the matrix multiplication technique. These formulae can in fact be easily derived using the triple method described here.

In the next chapter too there exist standard formulae for the various problems, but these bear no obvious relationship to each other. The triple method, on the other hand, offers a simple, easy device for solving them all.

5 COORDINATE GEOMETRY

LENGTH OF PERPENDICULAR

Example 1 Find the length of the perpendicular from the point (5,2) onto the line 4y = 3x.

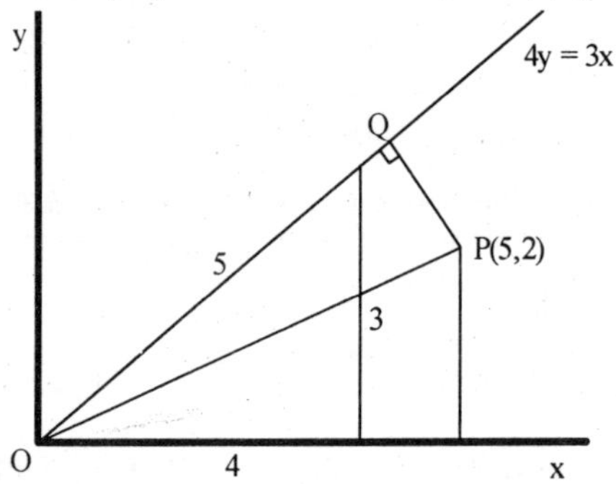

We require PQ in the diagram above.
Subtracting the triple for P from the triple for the line gives the triple for ΔOPQ:

line:	4	3	5	
P:	5	2	$\sqrt{29}$	–
	26	7	$5\sqrt{29}$	
∴ ΔOPQ =	$\frac{26}{5}$	$\frac{7}{5}$	$\sqrt{29}$	

∴ PQ = $\frac{7}{5}$ = 1.4.

Note:
1. we require only the middle element of the triple for ΔOPQ so we do not need to calculate the others,
2. subtracting the line from the point would give the same result but with the sign different, therefore we need not trouble ourselves about which triple to subtract from which.

Example 2 Find the distance of the point (–2,–5) from the line $5y + 12x = 0$.

This is the same as finding the length of the perpendicular.
Think of the equation of the line as $5y = -12x$ to get the line triple.

line:	5	–12	13	
P:	–2	–5	-	–
	-	49	-	

$\therefore$ Distance = $\frac{49}{13}$

Example 3 Find the distance of (6,3) from $3y = 4x + 2$.

Here the line does not pass through the origin.
But we observe that the point (1,2) lies on the line and transpose the origin to there. Then find the distance of (5,1) from $3y = 4x$:

3	4	5	
5	1	-	–
-	17	-	

$\therefore$ Distance = $\frac{17}{5}$ = 3.4.

No adjustment is required due to the transposition of the origin.

Example 4 Find the distance of $3x - 4y = 5$ from the origin.

We can transpose the origin to the point (3,1), which is on the line and then find the distance of (–3,–1) from $3x - 4y = 0$:

4	3	5	
–3	–1	-	–
-	–5	-	

$\therefore$ Distance = 1.

FOOT OF PERPENDICULAR

Example 5 Find the coordinates of the foot of the perpendicular from P(7,4) onto the line $y = 2x$.

We find the reflection P' (see previous chapter) of the point in the line and find the mid-point of PP' by *taking the average* of the x and y coordinates:

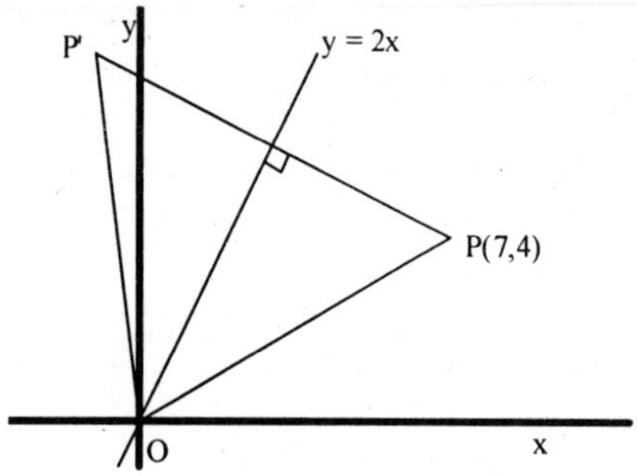

line:	1	2	$\sqrt{5}$
2 × line:	–3	4	5
P:	7	4	-
	–5	40	-
P' =	–1	8	-

$\therefore$ the foot of the perpendicular is at $\left(\frac{7-1}{2}, \frac{4+8}{2}\right) = (3,6)$.

This is not the most efficient way to solve this kind of problem though.

Exercise A Find the length of the perpendicular from:

1. (–3,2) onto 3y = 4x
2. (0,4) onto 24y = 7x
3. (9,10) onto 3y = 4x + 3
4. (3,5) onto 6y + 8x = 3
5. (1,1) onto 12y = 5x – 13
6. (4,–5) onto y = 3x – 4
7. (6,2½) onto 4y = 3x
8. (4,5) onto y = 2x + 3

ANGLE BETWEEN TWO LINES

Example 6 Find the angle between the lines 3y = 4x – 12 and y = x + 10.

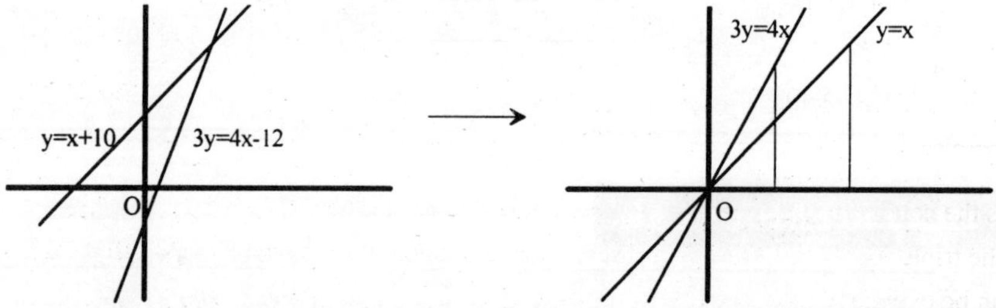

We are interested here only in the slope of the lines and so can disregard the absolute terms –12 and 10.

Then subtracting the line triples gives a triple for the angle between the two lines:

3	4	5	
1	1	$\sqrt{2}$	–
7	1	$5\sqrt{2}$	which represents the required angle.

Example 7 Find the angle between y = 2x + 9 and 2y = –x – 4.

1	2	$\sqrt{5}$	
2	-1	$\sqrt{5}$	–
0	5	5	= 0, 1, 1

The lines are therefore perpendicular.

What is the condition for lines to be perpendicular?
Let Ay = Bx and ay = bx be two lines, then since:

	A	B	-
	a	b	- –
=	(Aa + Bb),	(Ba – Ab),	- (1)

and since the triple for 90° is 0, 1, 1, then **Aa + Bb = 0** is the condition for lines to be perpendicular.

Thus in the above example: $1\times2 + 2\times(-1) = 0$.

Example 8 Find the angle between y = 2x + 1 and 3y = x.

1	2	$\sqrt{5}$	
3	1	$\sqrt{10}$	–
5	5	$\sqrt{50}$	= 1, 1, $\sqrt{2}$

The angle is therefore 45°.

What is the condition for lines to be inclined at 45° to each other?
Since the triple for 45° is 1,1, $\sqrt{2}$, from (1) above the condition must be: **Aa + Bb = Ba – Ab**.
This can be expressed as $\frac{A}{B} = \frac{a-b}{a+b}$

Thus in the above example: $\frac{1}{2} = \frac{3-1}{3+1}$.

Though angles between pairs of lines may be left in triple form, later we will see how these can be converted into an angle.

Exercise B Find the angle, in triple form, between the lines:

1. $y = 3x, y = 2x$
2. $2y = 3x, 3y = 4x$
3. $2y = 3x + 4, y = x + 3$
4. $y = x, 4y = 3x$
5. $3y = 4x, 4y = 3x$
6. $5y = 12x, 3y = 4x$
7. $2y = 3x, 5y = x$
8. $2y = 3x, 3y = 2x$
9. $3y = 2x, 2y + 3x = 0$
10. Find the equation of the angle bisector of the lines $3y = x$ and $31y = 27x$.

EQUATION OF A LINE

The following method for finding the equation of a straight line through two given points is a slightly modified version of that in the book "Vedic Mathematics" (Reference 1).

Example 9 Find the equation of the line through (6,5) and (4,1).

$$\begin{array}{cc} 6 & 5 \\ 4 & 1 \\ \hline \end{array}$$
$$\underline{(6-4)y = (5-1)x + (6\times1 - 5\times4)}$$

$\therefore\ 2y = 4x - 14$

The coefficient of y is found by subtracting the x-coordinates,
the coefficient of x is found by subtracting the y-coordinates
and the third term is the determinant: $\begin{vmatrix} 6 & 5 \\ 4 & 1 \end{vmatrix}$ i.e. $6\times1 - 5\times4 = -14$.

$\therefore\ \underline{y = 2x - 7}$

This is an application of the Vedic formula *Vertically and Crosswise.*

Example 10 Find the equation of the line through (–1,2) and (3,–5).

$$\begin{array}{cc} 3 & -5 \\ -1 & 2 \\ \hline \end{array}$$
$$\underline{4y = -7x + 1}$$

A proof of this is given in the proofs section at the end of this book.

Exercise C Find the equation of the line through:

1. (7,3) and (2,1)
2. (8,–1) and (2,–5)
3. (–1,7) and (–5,4)
4. (–3,0) and (1,6)
5. (0,7) and (–7,0)
6. (0,–3) and (–1,–2)

FURTHER EXAMPLES

Example 11 Find the equations of the lines through (5,2) which make an angle of 45° to the line $y = 2x + 3$.

There are two possibilities as the 45° could added to or subtracted from the slope of the line:

line:	1	2	-	
45°	1	1	-	+
	–1	3	-	

$\therefore\ -y = 3x + c$

line:	1	2	-	
45°	1	1	-	–
	3	1	-	

$\therefore\ 3y = x + c$

and since the lines pass through the point (5,2) we get:

$\underline{-y = 3x - 17}$ and $\underline{3y = x + 1}$.

Example 12 Find the coordinates of the points of contact of the tangents to the circle $x^2 + y^2 - 6x - 8y + 16 = 0$ which pass through the origin.

[The reader is invited to solve this example and the next by a non-triple method]

The centre of the circle is at (3,4) and the radius is 3.

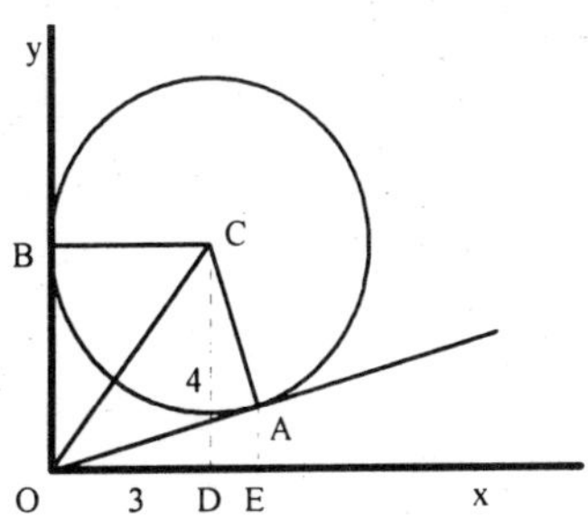

So OC = 5

and OA = OB = 4.

For A we obtain ΔOEA by subtracting ΔOAC from ΔODC:

	3	4	5	
	4	3	5	–
angle AOE	24	7	25	

$\therefore$ as OA = 4 the triple for angle AOE must be multiplied by $\frac{4}{25}$ i.e. $\frac{96}{25}, \frac{28}{25}, 4$.

$\therefore$ $A\left(\frac{96}{25}, \frac{28}{25}\right)$ And B is clearly at (0,4).

Example 13 Find the equation of the circle, centre (3,4), to which the line y = 2x is tangential.

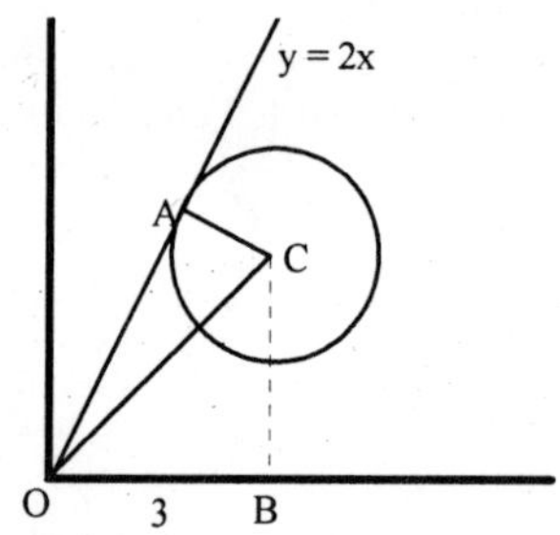

We need the radius of the circle, AC.
Since the sum of the two triangles shown has a gradient of 2 we can find ΔOAC by triple subtraction:

	1	2	$\sqrt{5}$	
	3	4	5	–
angle AOC	11	2	$5\sqrt{5}$	$= \frac{11}{\sqrt{5}}, \frac{2}{\sqrt{5}}, 5$, since OC = 5.

$\therefore$ $AC = \frac{2}{\sqrt{5}}$ and the equation of the circle is $(x-3)^2 + (y-4)^2 = \frac{4}{5}$.

The problem below is from an NEAB (UK examination board) exam paper (1990, Paper 1).

✱ Tangents are drawn from the point A(–2,6) to meet the circle $x^2 + y^2 = 8$ at B and D, where B is in the first quadrant. Show B(2,2) and find the coordinates of D. Find also the angle BAD.

We can subtract the triple for angle BOA (where O is the origin) from the point triple (i.e. –1,3, $\sqrt{10}$ – 1,2, $\sqrt{5}$) to get a triple for the point B, OB being known.
For D we add –1,3, $\sqrt{10}$ and 1,2, $\sqrt{5}$ (answer –2.8,0.4).
Angle BAD is the angle complementary to 1,2, $\sqrt{5}$, which is then doubled (answer 3,4,5).

6 TRIPLE CODE NUMBERS

GROUPING TRIPLES

We return now to the triples themselves and look at the patterns and groupings among them. If we look for the perfect triples all of whose elements are between 1 and 100 and where the first element is greater than the second we obtain the following:

4	3	5	63	16	65
12	5	13	21	20	29
24	7	25	45	28	53
15	8	17	56	33	65
40	9	41	77	36	85
60	11	61	80	39	89
35	12	37	55	48	73
84	13	85	72	65	97

Studying these 16 triples we may observe three groups:

	Group 1	
4	3	5
12	5	13
24	7	25
40	9	41
60	11	61
84	13	85

	Group 2	
15	8	17
35	12	37
63	16	65

	Group 3	
21	20	29
45	28	53
56	33	65
77	36	85
80	39	89
55	48	73
72	65	97

Referring to the three elements in each triple by x, y, r we see in Group 1:

(a) y = the odd numbers in order from 3 onwards,

(b) $y^2 = x + r$,

(c) $r = x + 1$.

Using these patterns we can extend the list of triples in Group 1:
from (a) above, the next middle number must be 15,
from (b) $15^2 = 225$ = the sum of the outer numbers
and from (c) we split 225 into two consecutive numbers to get 112 and 113.

So the next entry in this list must be 112, 15, 113: which we may verify is also a triple.

Continuing we get:			
	112	15	113
	144	17	145
	180	19	181
	220	21	221
		etc.	

In Group 2 we see: (a) y = multiples of 4 starting at 8,
(b) $\frac{y^2}{2} = x + r$,
(c) $r = x + 2$.

Again we can extend the list of triples in Group 2:
from (a) the next middle number must be 20,
from (b) $\frac{20^2}{2} = 200$ = the sum of the outer numbers
and from (c) the outer numbers must be 99 and 101.

So the next triple is	99	20	101	and continuing:
	143	24	145	
	195	28	197	
	255	32	257	
		etc.		

Group 3 we leave for the moment as it has no obvious patterns.

CODE NUMBERS

We now bring groups 1 and 2 together:

4	**3**	5
15	**4**×2	17
12	**5**	13
35	**6**×2	37
24	**7**	25
63	**8**×2	65
	etc.	

TABLE 1

Here the triples in the two groups alternate and the even numbers in Group 2 are written as 4 × 2, 6 × 2 etc. so that the series of whole numbers from 3 onwards appears in a vertical column (in bold in Table 1).

The numbers in this column we call the **code numbers** of the triples:
3 is the code number for the triple 4, 3, 5;
4 is the code number for 15, 8, 17 and so on.

Thus each code number represents a triple and every triple in the two groups fits in and has its code number.
We will denote the code number in general by the letter c.

We see that for those triples in which the middle element is odd, that number is the triple code number; and for those in which the middle element is even the code number is half that even number.

Example 1 For the triple 684, 37, 685 the code number is 37, i.e. c = 37.

Example 2 For the triple 1443, 76, 1445 the code number is 38, i.e. c = 38.

Example 3 Find the triple with code number 19.

Since 19 is odd, square it to get 361 and find two consecutive numbers which add up to 361.
∴ the triple is 180, 19, 181.

Example 4 Find the triple for which c = 22.

Since 22 is even, square it to get 484 and put down one less and one more than this for the outer elements of the triple. Then put double the code number for the middle element.
∴ the triple is 483, 44, 485.

Exercise A

Write down the code number of the following triples:

1. 420, 29, 421 2. 1763, 84, 1765 3. 3120, 79, 3121 4. 4095, 128, 4097

Find the triples with the following code numbers:

5. 32 6. 37 7. 28 8. 101

GEOMETRICAL SIGNIFICANCE OF THE CODE NUMBERS

Let us take a vertical unit line and complete a triangle having this as height and 3 units as base:

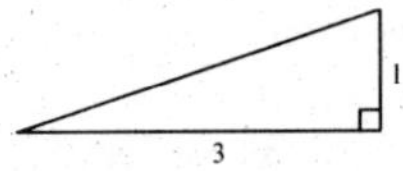

Let this triangle be ½A) $3, 1, \sqrt{10}$.

Then doubling the angle (see Chapter 2) we have A) 8, 6, 10 = 4, 3, 5.
These two triangles are shown below:

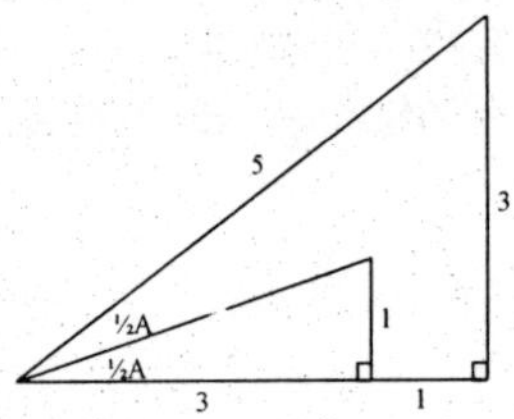

If we now take the same vertical unit with a base of 4, we can similarly produce a triangle with double the angle:

½B) 4 1 $\sqrt{17}$
B) 15 8 17

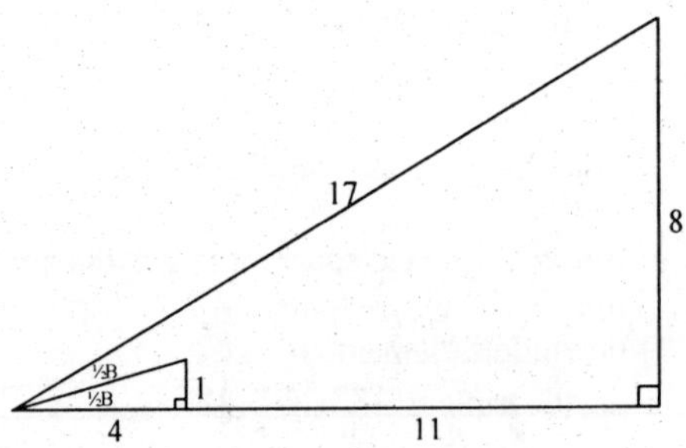

We can continue this process, adding 1 to the base each time:

½C) 5 1 $\sqrt{26}$
C) 24 10 26
= 12 5 13

½D) 6 1 $\sqrt{37}$
D) 35 12 37

½E) 7 1 $\sqrt{50}$
E) 48 14 50
= 24 7 25

We can see that we are producing the triples in Table 1.
We also see that the length of the base line that we put with the unit vertical is the code number of the triple: for A) 4,3,5 the code number was 3 and this is the base of the half angle triangle as shown above.
We will write this as c(A) = 3, meaning that the code number of the triple containing the angle A is 3. Similarly c(B) = 4, c(C) = 5, c(D) = 6, c(E) = 7 and so on.

CODE NUMBER PAIRS

Let us now take a unit vertical and a base of 2½.
Since this triangle will still contain the same angle if both sides are doubled

we can use 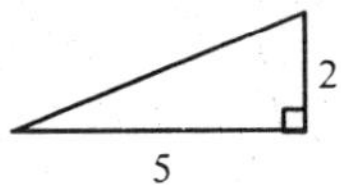 instead of 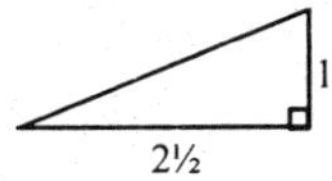 if we wish.

So if ½A) 5 2 $\sqrt{29}$
then A) 21 20 29 and we obtain the first triple in Group 3.

In fact the triples in Group 3 are all produced, like this one, from half angle triangles where the base length is a rational number which is not a whole number, when the height is 1.

Such triples can therefore be described by a pair of code numbers: 5, 2 in this case.

We will use c, d for pairs of code numbers, so c, d = 5, 2 in the example above.

Let us next examine the case where c, d = 7, 2:

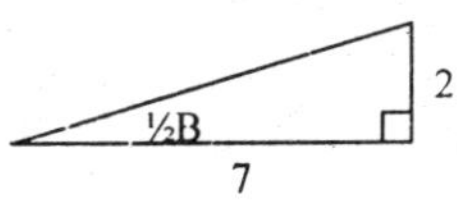

½B) 7 2 $\sqrt{53}$
∴ B) 45 28 53

this triple is the second in Group 3.

It will be seen that a) triples with a single code number, as in the previous section, could be written as a pair: c = 3 is the same as c, d = 3, 1,
b) triples with a pair of code numbers can be written with a single code number: c, d = 5, 2 is the same as c = 2½.

CODE NUMBERS AS TRIPLES

Because code numbers represent the sides of a triangle, they can themselves form a triple.

E.g. for c = 6 the triple is 6, 1, $\sqrt{37}$.
And for c, d = 7, 3 the triple is 7, 3, $\sqrt{58}$.

This means that the code numbers can be divided through just like triples (we were in fact doing this in the last section). For example, 14, 6 = 7, 3
and 9, 3 = 3, 1 = 3.

If, instead of seeing the code number triangle as having half the angle of the triple it generates, we see the triple as having double the angle of its code number triangle, we have the following useful relationship: the first two elements of a triple are the code numbers of a triple with double the angle.

That is, for the triple A) x, y, z $c(2A) = \frac{x}{y}$

We will find this result very useful later on.

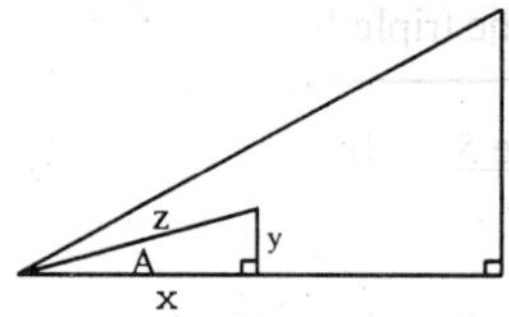

ALGEBRAIC FORMULATION

Consider the general triple: ½A) c d $\sqrt{c^2+d^2}$.

Doubling this we get: A) $c^2 - d^2$ 2cd $c^2 + d^2$.

That is, diagrammatically:

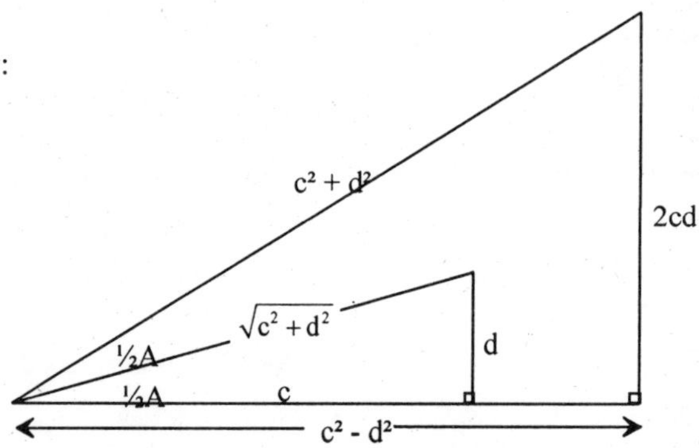

From this we note the following:

1. If c and d are rational the triple for A will have rational elements.

2. $c^2 - d^2$, 2cd, $c^2 + d^2$ is a formula for generating rational (i.e. perfect) triples.
   That is, for all rational values of c and d we can generate a perfect triple. c, d are the code numbers of the triples as before.
   [It can also be shown that this formula generates all possible perfect triples.]

3. If d = 1 we have A) $c^2 - 1$, 2c, $c^2 + 1$ and putting c = 3, 4, 5, 6 etc. we get, in turn, each of the triples in Table 1:

   for c = 3 we obtain 8, 6, 10 = 4, 3, 5,
   for c = 4 we obtain 15, 8, 17,
   etc.

CONVERTING CODE NUMBERS TO TRIPLES

To get the triple from its code numbers we can use **$c^2 - d^2$, $2cd$, $c^2 + d^2$.**

Example 5 If c, d = 10, 3 the triple is $10^2 - 3^2$, $2\times10\times3$, $10^2 + 3^2 = 91, 60, 109$.

Example 6 If c = 6 the triple is $6^2 - 1$, 2×6, $6^2 + 1 = 35, 12, 37$.

Example 7 If c, d = –6, –4 then c, d = 3, 2 and the triple is 5, 12, 13.

CONVERTING TRIPLES TO CODE NUMBERS

If the triple is recognized as being in Table 1 then the code number can be found as explained earlier. Otherwise there are two methods.

Method A. *By Addition and By Subtraction*: add the outer pair of the triple and take the square root. This gives c. Subtract the outer pair and take the square root. This gives d.

Example 8 Find the code numbers for 21, 20, 29.

$$c, d = \sqrt{29+21}, \sqrt{29-21} = \sqrt{50}, \sqrt{8} = \sqrt{25}, \sqrt{4} = \underline{5, 2}.$$

Proof: $c^2 - d^2$, $2cd$, $c^2 + d^2$ is a triple with code numbers c and d.
And $\sqrt{(c^2+d^2)+(c^2-d^2)}, \sqrt{(c^2+d^2)-(c^2-d^2)} = \sqrt{2c^2}, \sqrt{2d^2} = c, d$.

However because two values arise when taking the square roots and because the middle element, which may be positive or negative, is not taken into account, this method should be confined to cases where all the elements of a triple are positive (unless consideration is given to the quadrant in which the triple lies).

Method B: By taking the sum of the outer elements of the triple, and the middle element.

Example 9 For 21, 20, 29, c, d = (21 + 29), 20 = $\underline{5, 2}$ as before.

Proof: $c^2 - d^2$, $2cd$, $c^2 + d^2$ has code numbers c and d.
And $(c^2 - d^2 + c^2 + d^2)$, $2cd = 2c^2$, $2cd$ = c, d.

This result has also been proved geometrically in showing how to find a half-angle triple.

No difficulties arise in using this method and it is also easier to calculate than the other method.

Exercise B Obtain triples for each of the following code numbers:

1. 10 (i.e. c=10) 2. 20 3. 100 4. 8, 3 5. 4, 3

6. 9, 2 7. 7, 3 8. 11, 4 9. 13, 7 10. 11, 9

11. 4, ½ 12. –7, –3 13. –1, 2 14. –3, 2 15. $\frac{1}{3}$

Obtain the code numbers for the following triples:

16. 12, 35, 37 17. 28, 45, 53 18. 45, 28, 53 19. 68, 285, 293

20. 140, 51, 149 21. 60, 91, 109 22. –16, 63, 65 23. 117, –44, 125

24. –33, –56, 65 25. 319, –360, 481

ADDITION AND SUBTRACTION OF CODE NUMBERS

Because a pair of code numbers themselves represent a triangle or triple, code numbers can be added and subtracted in the same way as triples.

Example 10 Add the triples A) 21, 20, 29 and B) 4, 3, 5.

By the usual method:

A	21	20	29	
B	4	3	5	+
A+B	24	143	145	

Or we can add the code numbers of these triples and convert this to a triple.
The code numbers for the above triples are 5, 2 and 3, 1:

c(A)	5	2	
c(B)	3	1	+
c(A+B)	13	11	

where 13 = 5×3 – 2×1 and 11 = 2×3 + 5×1
That is, c(A) + c(B) = c(A+B). *

We then convert 13, 11 to the triple 24, 143, 145.

The arithmetic is easier here because the code numbers are always smaller numbers than the numbers in the triple itself.
This is particularly useful when we only want the code numbers of the answer anyway, as we will see later.

The code number sum above can be written as 5, 2 + 3, 1 = 13, 11
or as $\frac{5}{2}+3=\frac{13}{11}$.

In order not to confuse triple addition with ordinary addition of numbers we will use a special symbol for triple addition: a plus sign with a dot over it ($\dot{+}$).

So $\frac{5}{2} \dot{+} 3 = \frac{13}{11}$.

Similarly we can use a minus sign with a dot over it ($\dot{-}$) for triple subtraction.

This notation will be used in Chapter 9.

* So the statement c(A) + c(B) = c(A+B) from the last example can be written more consistently as c(A) $\dot{+}$ c(B) = c(A+B).

Example 11 Find the code numbers for 8, 15, 17 $\dot{-}$ 77, 36, 85.

The code numbers for these triples are 5, 3 and 9, 2.

$$\begin{array}{cc} 5 & 3 \\ \underline{9} & \underline{2} \end{array} \dot{-}$$
$$\underline{51 \quad 17} = \underline{3 \quad 1}.$$

CODE NUMBERS OF CODE NUMBERS

Example 12 The code numbers of 4, 3, 5 are 3, 1 which, as a triple is 3, 1, $\sqrt{10}$.

And the code numbers for 3, 1, $\sqrt{10}$ are 3+$\sqrt{10}$,1.

Example 13 For 119, 120, 169, c, d = 12, 5;
for 12, 5, 13, c, d = 5, 1;
for 5, 1, $\sqrt{26}$, c, d = 5+$\sqrt{26}$, 1;
etc.

CODE NUMBERS OF COMPLEMENTARY TRIPLES (CT)

Complementary triples add up to 90° and we transpose the first two elements of a triple to obtain its complementary triple.

Example 14 If c, d = 20, 9 then *By Addition and By Subtraction:*

c, d(CT) (i.e. the code numbers of the complementary triple) = 20+9, 20-9
= 29, 11.

Example 15 And if c, d = 29, 11 then c, d(CT) = 29+11, 29–11 = 40, 18 = 20, 9.

Example 16 If c, d = 5, –2 then c, d(CT) = 3, 7.

Proof: For the triple $c^2 - d^2$, 2cd, $c^2 + d^2$ with code numbers c and d the complementary triple is 2cd, $c^2 - d^2$, $c^2 + d^2$ (see Chapter 1).
Then by the method used in Example 8 the code numbers of this triple are:
c, d(CT) = $\sqrt{c^2 + d^2 + 2cd}, \sqrt{c^2 + d^2 - 2cd}$ = (c + d),(c – d).

❋ Use the method of Example 9 to show that the code numbers of 2cd, $c^2 - d^2$, $c^2 + d^2$ are c+d, c–d.

Since the sum of two complementary triples is 90° (i.e. the triple 0, 1, 1) the sum of their code numbers is 45° (triple 1, 1, $\sqrt{2}$).
For verification we can therefore add 20, 9 and 29, 11 in Example 14 and expect to get 1, 1.

20	9			
29	11	+		
481	481	=	1, 1	as expected.

CODE NUMBERS OF SUPPLEMENTARY TRIPLES (ST)

Supplementary triples add up to 180° (i.e. –1, 0, 1) and we change the sign of the first element of a triple to obtain its supplementary triple (see Chapter 2).

Example 17 If c, d = 20, 9 then *by transposition:*

c, d(ST) (i.e. the code numbers of the supplementary triple) = 9, 20.

Example 18 If c = 7 then c, d(ST) = 1, 7.

Proof: For $c^2 - d^2$, $2cd$, $c^2 + d^2$ which has code numbers c, d the supplementary triple is $d^2 - c^2$, $2cd$, $c^2 + d^2$

$\therefore$ c, d(ST) = $(d^2 - c^2 + c^2 + d^2)$, 2cd = d, c.

Again, for verification, supplementary triples add up to –1, 0, 1 (180°) so that their code numbers add up to 0, 1, 1 (90°).

Using Example 17:

$$\begin{array}{rr} 20 & 9 \\ \underline{9} & \underline{20} \; + \\ \underline{0} & \underline{481} \end{array} = \underline{0,\ 1} \text{ as expected.}$$

CODE NUMBERS FOR 0°, 90°, 180°, 270°

These are summarised below:

Angle	**0°**	**90°**	**180°**	**270°**
Triple	1, 0, 1	0, 1, 1	–1, 0, 1	0, –1, 1
c, d	1, 0	1, 1	0, 1	1, –1
c	infinity	1	0	–1

In the triangle opposite the number in each angle denotes the triple code number of that angle.

So the angle marked c=1 is a right angle (see above).

It follows that the other angles must add up to 90°, which means that their code numbers must add up to 1:

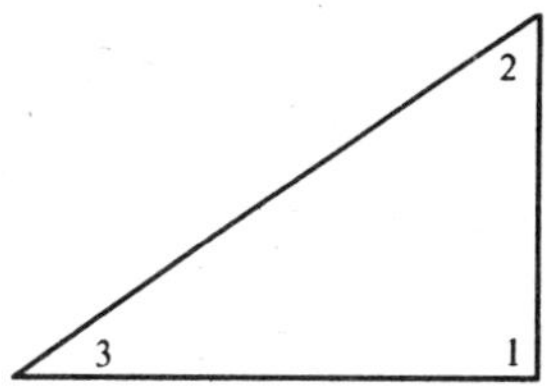

$$\begin{array}{rr} 3 & 1 \\ \underline{2} & \underline{1} \; + \\ \underline{5} & \underline{5} \end{array} = 1,\ 1 = 1 \text{ as predicted (i.e. } 3 + 2 = 1\text{).}$$

In fact this is the familiar 4, 3, 5 triangle.

✱ Confirm that $\frac{3}{7}$, 3, 12 can be the code numbers of the angles of a triangle.

RELATION BETWEEN CODE NUMBERS AND ANGLES

How do the code numbers change as the angle increases from 0° to 360°?
Consider the eight triples shown together diagrammatically in Chapter 2 and their code numbers:

4	3	5	c=3
3	4	5	2
–3	4	5	$\frac{1}{2}$
–4	3	5	$\frac{1}{3}$
–4	–3	5	$-\frac{1}{3}$
–3	–4	5	$-\frac{1}{2}$
3	–4	5	–2
4	–3	5	–3

We can now plot these triangles and their code numbers, together with those for the quadrant boundaries (and 4 other triangles), on a single diagram:

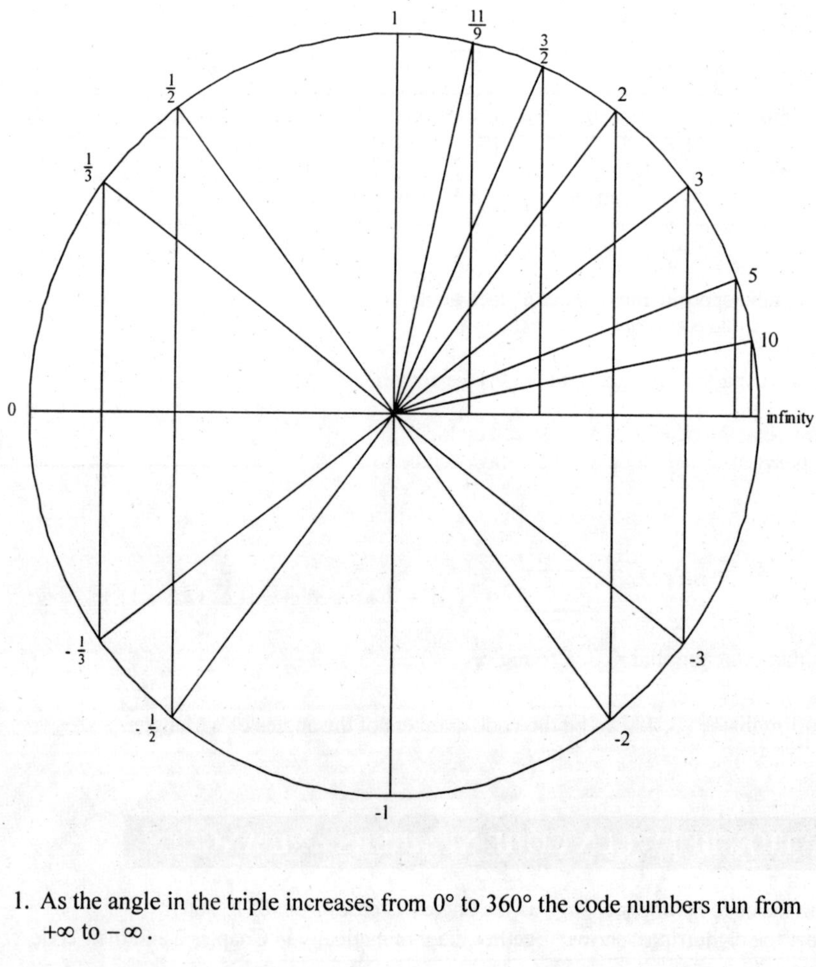

Note: 1. As the angle in the triple increases from 0° to 360° the code numbers run from $+\infty$ to $-\infty$.

2. There is symmetry about the horizontal line, i.e. the code numbers below are minus those above. So $c(-A) = -c(A)$.

3. There is symmetry about the vertical line, i.e. the code numbers on the left are the reciprocal of those on the right. So $c(180° - A) = \frac{1}{c(A)}$.

These relationships will be found useful later on.

Example 19 In which quadrant does the triple with code numbers –3, 4 lie?

Since $c = \frac{-3}{4}$ which lies between –1 and 0, the triple is in the third quadrant.

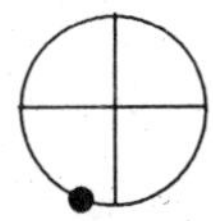

Example 20 To which quadrant boundary is the triple with code numbers 1, 41 nearest?

Since $c = \frac{1}{41}$ which is a small positive quantity the triple is close to –1, 0, 1 (180°) and is in the second quadrant.

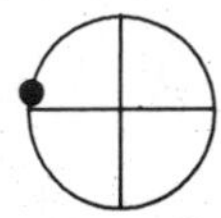

FURTHER EXAMPLES

Example 21 Find the equation of a circle of radius 4 which touches the x-axis and the line 4y = 3x.

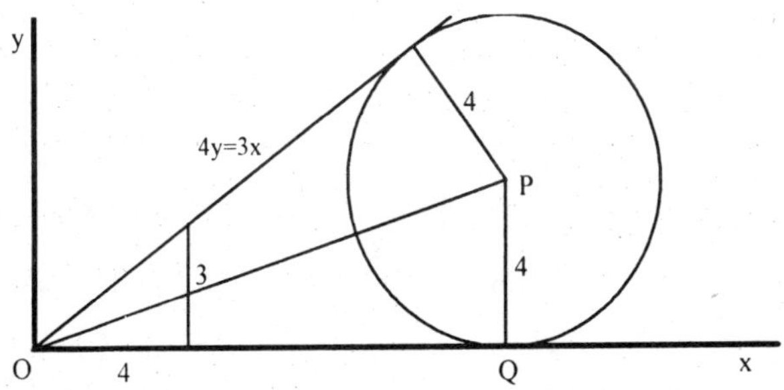

The line triple is 4,3,5 with code numbers 3,1.
∴ as OP bisects the angle formed by the two tangents, ΔOPQ is represented by the triple 3, 1, -.

But since PQ=4 the coordinates of P are (12,4).

The equation of the circle is then $\underline{(x - 12)^2 + (y - 4)^2 = 4^2}$.

Example 22 Find the coordinates of the centre of a circle, radius 6, which touches the lines $3y = 4x$ and $12y = 5x$.

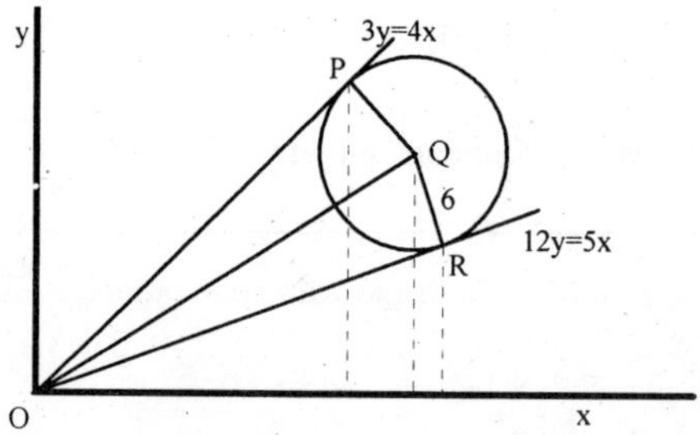

By subtracting the triples for the two given lines we get a triple for angle POR.
The code numbers then give us a triple for angle QOR.
We can then add this to the triple for $12y = 5x$ to get a triple for angle QOx.

So:

$$\begin{array}{ccc} 3 & 4 & 5 \\ 12 & 5 & 13 \\ \hline 56 & 33 & 65 \end{array} - \quad \text{triple for } P\hat{O}R$$

This has code numbers 11, 3 so **11,3,-** is a triple for $Q\hat{O}R$.
And since QR = 6, OR = 22.

$$\therefore \quad \begin{array}{ccc} 22 & 6 & OQ \\ 12 & 5 & 13 \\ \hline 234 & 182 & 13OQ \end{array} + \quad = 18, 14, OQ.$$

The coordinates of Q are therefore (18,14).

Exercise C Find the code numbers of the sum or difference for the following triples:

1. 12, 5, 13 + 8, 15, 17
2. 39, 80, 89 + 24, 7, 25
3. 20, 21, 29 + 15, 8, 17
4. 55, 48, 73 – 77, 36, 85
5. 12, 5, 13 – 8, 15, 17

Find the code numbers of the triples complementary to the following:

6. 15, 8, 17
7. 80, 39, 89
8. 39, 80, 89
9. 20, 99, 101
10. 55, 48, 73
11. –24, 7, 25
12. 15, –8, 17
13. –28, 45, 53
14. –51, –140, 149

Find the code numbers of the triples supplementary to the following:

15. 12, 5, 13   16. –24, 7, 25   17. 3, –4, 5

Name the quadrant (1, 2, 3 or 4) in which the triple represented by each of the following code numbers lies:

18. 7, 4   19. 100, 3   20. 4,7   21. ½, 1

22. –10, 1   23. –10, –9   24. 10, –1   25. –2, 3   26. 9, –10

Name the quadrant boundary (0°, 90°, 180° or 270°) which the triple for each of the following code numbers is near:

27. 50, 1   28. 50, –1   29. 1, 50   30. 49, 51

31. –9, 8   32. –2, 101

SUMMARY

Here we summarise the most important results of this chapter.

1. The triple with code numbers **c, d** is $\mathbf{c^2 - d^2, 2cd, c^2 + d^2}$.

2. The code numbers of the triple **x, y, r** are **x+r, y.**

3. The triple **c, d, -** has half the angle of the triple $\mathbf{c^2 - d^2, 2cd, c^2 + d^2}$.
   Or, to put it another way, the first two elements of a triple are the code numbers of the double angle triple.

4. To find the complement or supplement of a triple or code number:

	Complement	**Supplement**
Triple	*transpose* the first two elements	change the sign of the first element
Code Numbers	*by addition and by subtraction*	*transpose*

7 SOLUTION OF TRIANGLES

In this chapter we consider the solution of plane, scalene triangles. The angles will be described in triple form. Conversion of angles to triples and triples to angles is explained in Chapter 9.

We will use the usual notation for triangles: A, B, C denote the three vertices and angles of a triangle and a, b, c will denote the lengths of the sides opposite those vertices (see diagram below).

It is worth noting that scalene triangles can always be solved using right-angled triangles though they are not usually solved that way.

There are three useful formulae:

the **angle-deficiency formula** which corresponds to the well-known cosine rule:

$$a^2 = b^2 + c^2 - 2bc\cos A;$$

The **sine formula**, which is equivalent to the conventional sine rule:

$$\frac{a}{\sin A} = \frac{b}{\sin B} = \frac{c}{\sin C}$$

and the **code-number formula** which corresponds to: $\dfrac{\tan\frac{1}{2}(A+B)}{\tan\frac{1}{2}(A-B)} = \dfrac{a+b}{a-b}$.

THE ANGLE-DEFICIENCY FORMULA

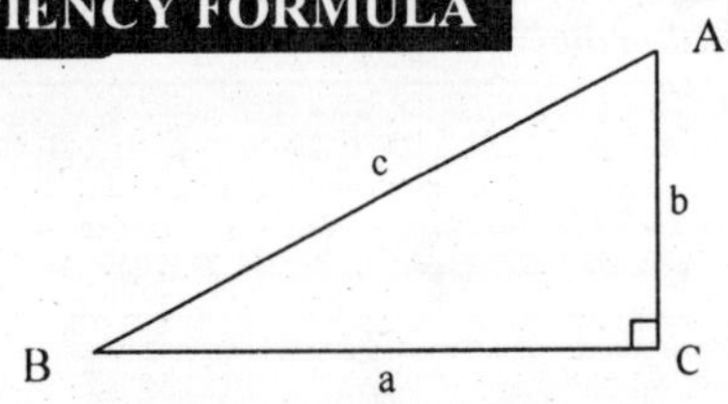

We know that in a right-angled triangle $a^2 + b^2 = c^2$ so that $a^2 + b^2 - c^2 = 0$.

And if $a^2 + b^2 - c^2 \neq 0$ then $\hat{C}$ is not a right angle.

In fact if $a^2 + b^2 - c^2 > 0$ then $\hat{C} < 90°$ i.e. acute;

and if $a^2 + b^2 - c^2 < 0$ then $\hat{C} > 90°$ i.e. obtuse.

In either case if the angle is not a right angle we will say it is deficient and we will call the quantity $\mathbf{a^2 + b^2 - c^2}$ the **angle deficiency** for $\hat{C}$.

So the sum of the squares of the sides enclosing an angle minus the square of the side opposite the angle is the angle deficiency of that angle.

<u>Example 1</u> Is a triangle with sides 2, 3, 4 an obtuse-angled triangle?

If the triangle is obtuse-angled the obtuse angle will be opposite the longest side.
We therefore find the angle deficiency for the angle opposite the side of length 4:
$2^2 + 3^2 - 4^2 = -3$ and this negative value indicates that the triangle is obtuse-angled.

<u>Example 2</u>

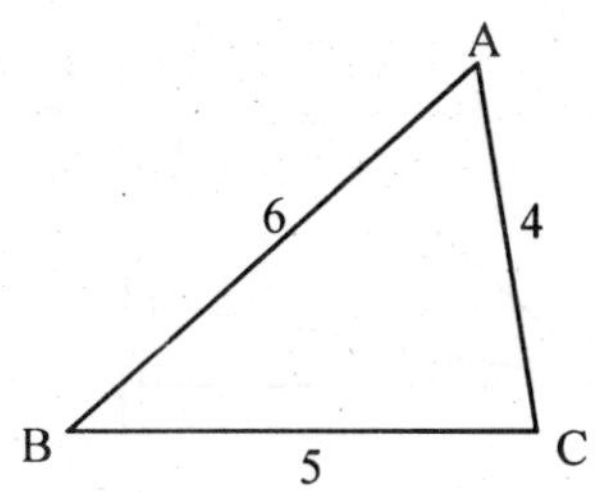

We can find the angle deficiencies for the angles A, B and C:

for A the deficiency is $6^2 + 4^2 - 5^2 = 27$
for B the deficiency is $5^2 + 6^2 - 4^2 = 45$
for C the deficiency is $4^2 + 5^2 - 6^2 = 5$.

This indicates that $\hat{C}$ is close to a right angle and acute.

We now drop a perpendicular AD from A onto BC:

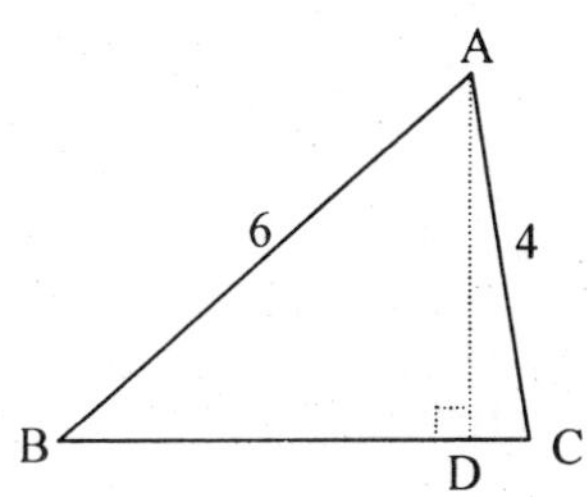

The angle-deficiency formula states that:

the angle-deficiency = the partial base × twice the base

where , if BC above is taken as base, the partial base can be DB, in which case we use the angle B, or DC, in which case we use the angle C.

<u>Example 3</u> Find DB and DC in the triangle above.

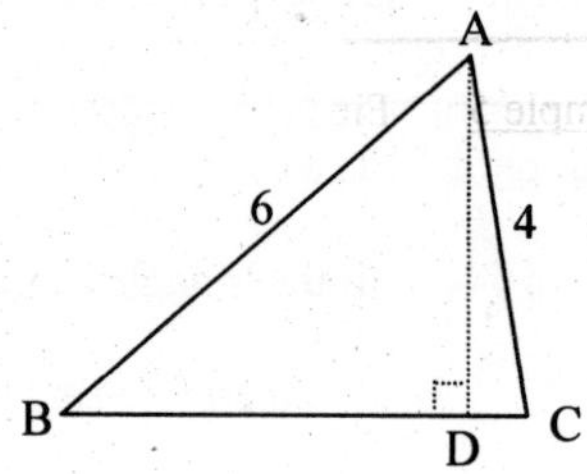

Since partial base = $\dfrac{\text{angle deficiency}}{\text{twice the base}}$

$\therefore$ DB = $\dfrac{\text{angle deficiency for B}}{2\times BC}$ $\underline{4.5}$.

And DC = $\dfrac{\text{angle deficiency for C}}{2\times BC} = \underline{0.5}$.

<u>Example 4</u> Similarly if we drop a perpendicular BE from B onto AC:

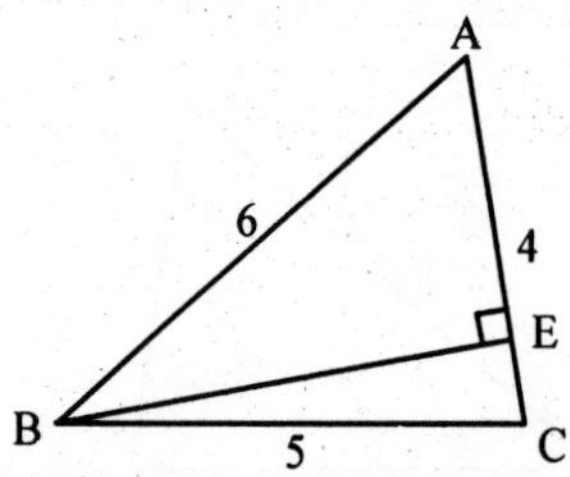

EC = $\frac{5}{8}$ and EA = $\frac{27}{8}$

Proof of the angle-deficiency formula:

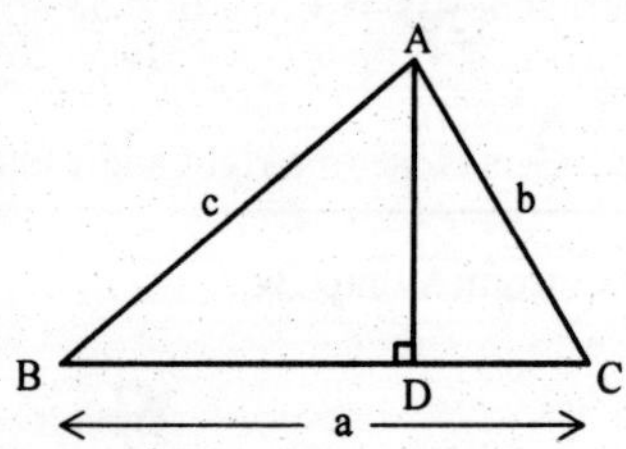

From the triple theorem: $AD^2 = b^2 - DC^2$

and $AD^2 = c^2 - (a - DC)^2$

$\therefore\ b^2 - DC^2 = c^2 - (a - DC)^2$

$\therefore\ b^2 = c^2 - a^2 + 2a\times DC$

$\therefore\ \mathbf{a^2 + b^2 - c^2 = DC\times 2a}$

If angle C is obtuse:

$AD^2 = b^2 - DC^2 = c^2 \ \ (a + DC)^2$

so that $a^2 + b^2 - c^2 = -DC \times 2a$.

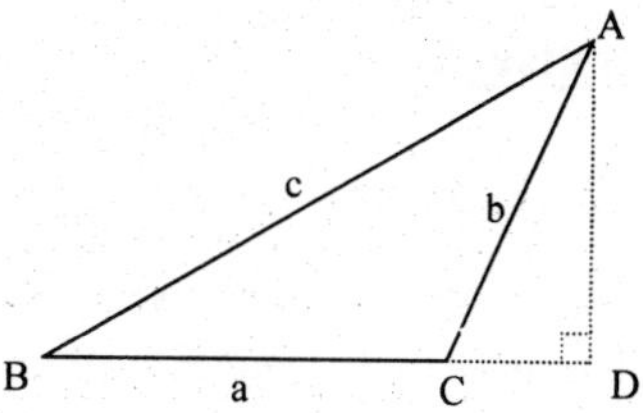

This shows that if the formula in bold above is used for obtuse-angled triangles DC will have a negative value.

Example 5 Find a triple for $\hat{B}$

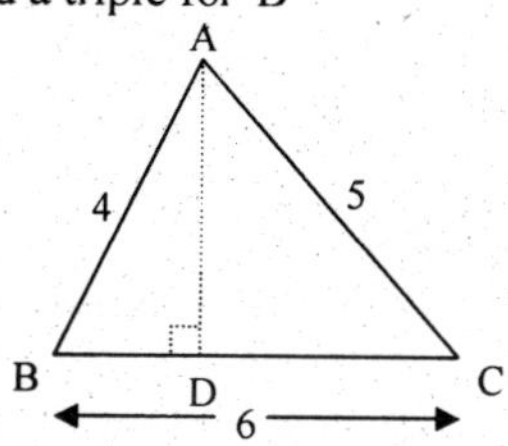

$BD = \dfrac{\text{angle deficiency for B}}{2 \times BC}$

$= \dfrac{4^2+6^2-5^2}{2\times 6} = \dfrac{9}{4}$

$\therefore$ B) $\frac{9}{4}$, - , 4 = 9, - , 16

Example 6 Find AC.

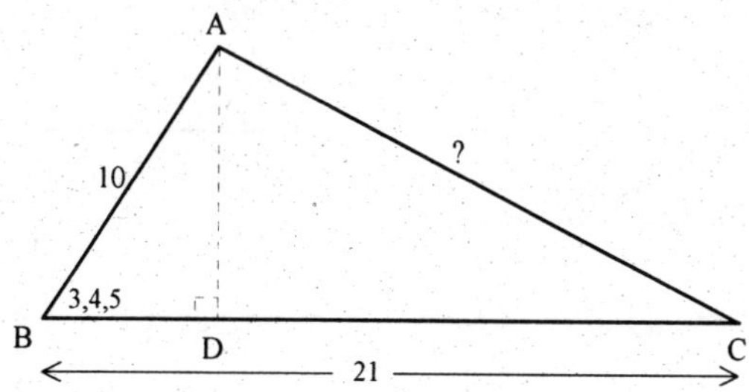

$\hat{B}$ is given by the triple 3,4,5 so that BD = 6.

Then the formula gives the angle-deficiency for B:
$10^2 + 21^2 - AC^2 = 6 \times 2 \times 21$.
$\therefore\ AC^2 = 541 - 252$.
$\therefore$ AC = 17.

Example 7 Find AC.

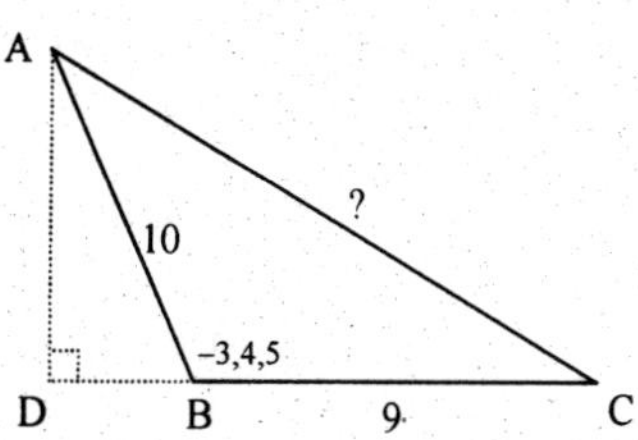

Similarly $BD = 10 \times \frac{-3}{5} = -6$.

$\therefore\ 10^2 + 9^2 - AC^2 = -6 \times 2 \times 9$.

$\therefore$ AC = 17

Exercise A

1. Find the six partial bases in the diagram opposite.

2. Find triples for the angles A, B and C in the diagram opposite

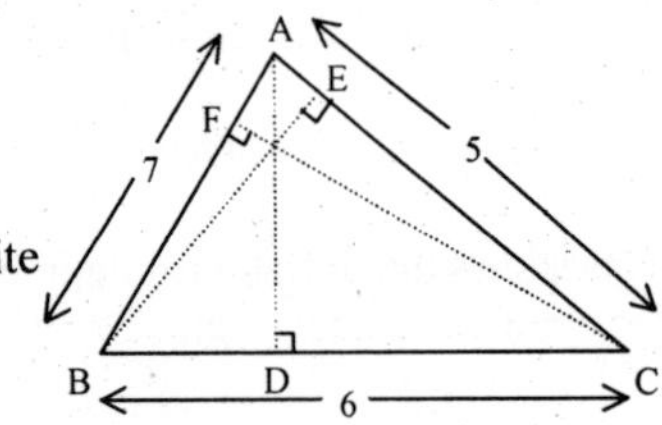

3. Find angle C:

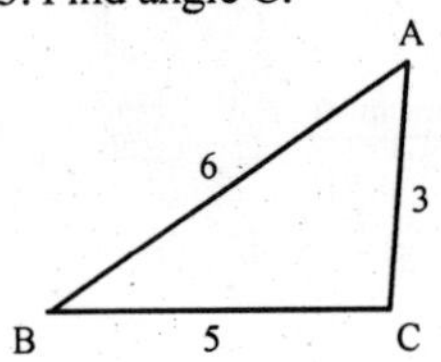

4. Find x

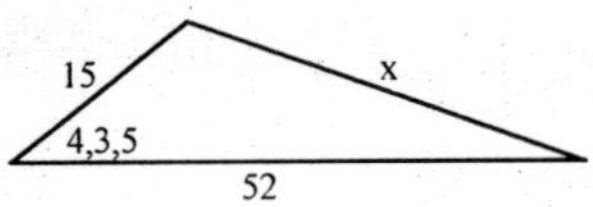

5. Find x

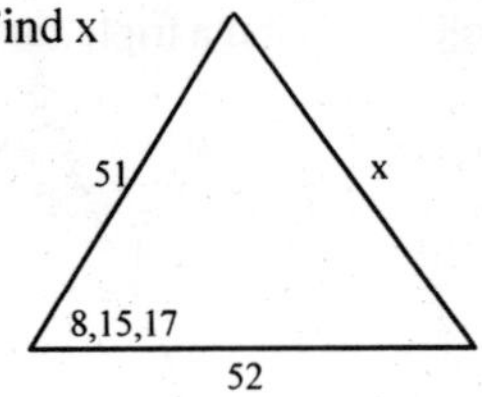

6. Find x

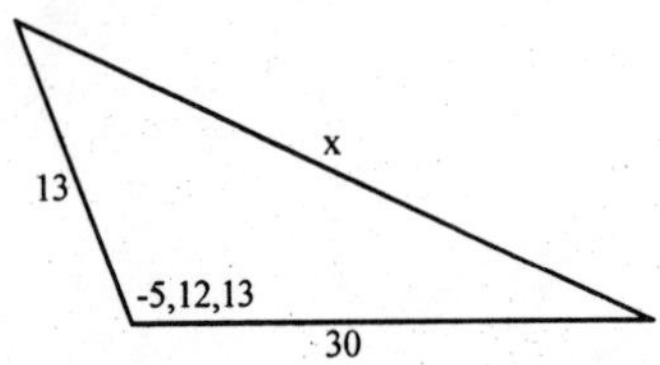

7. Find x

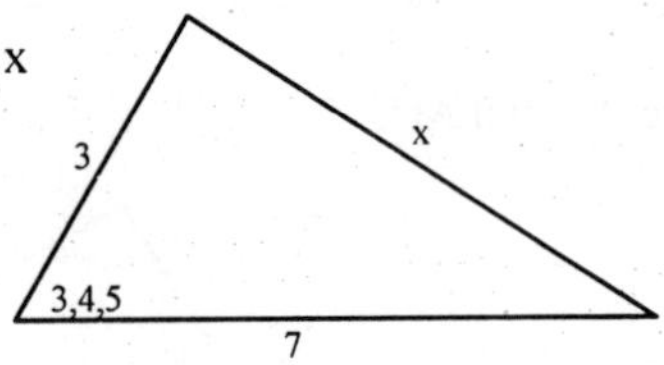

8. In ΔABC, AB = AC = 5cm and BC = 6 cm. A perpendicular CP is dropped from C onto AB. Find AP.

THE SINE FORMULA

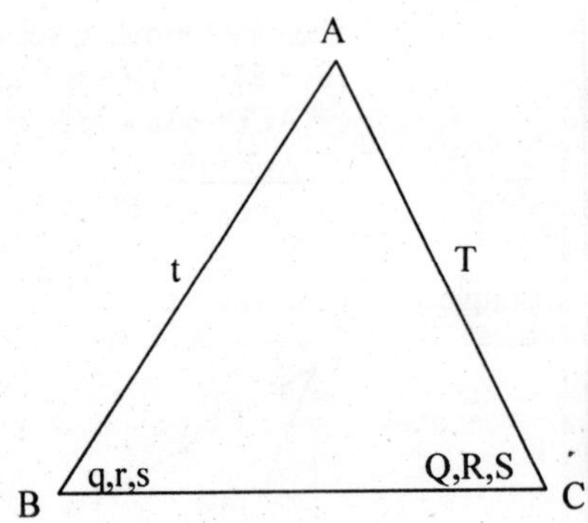

In the triangle ABC where AB = t and AC = T:

B) q r s
and C) Q R S

$\frac{tr}{s} = \frac{TR}{S}$ this is the sine formula (in triple form).

Proof: from the figure we see:

$AD = t\frac{r}{s}$ and $AD = T\frac{R}{S}$

$\therefore \frac{tr}{s} = \frac{TR}{S}.$

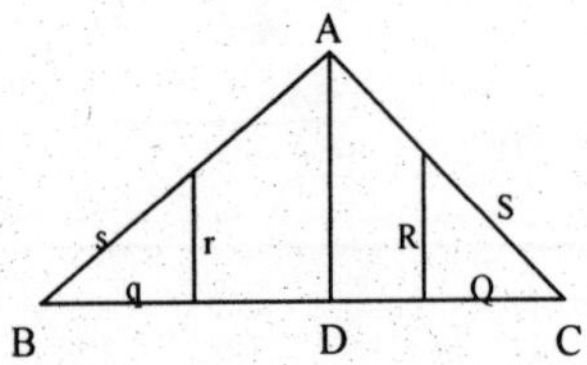

Example 8 Find x.

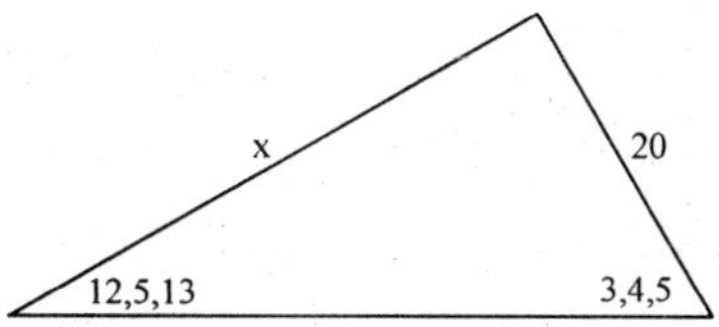

Using the sine formula:

$$\frac{x5}{13}=\frac{20\times4}{5}$$

$$\therefore x=\frac{16\times13}{5}=\underline{41.6}.$$

Example 9 Find x.

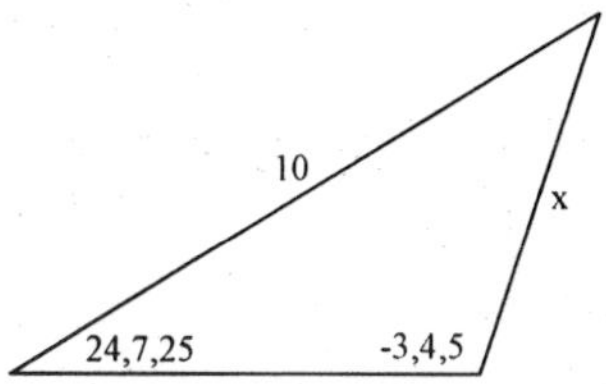

$$\frac{x4}{5}=\frac{10\times7}{25}$$

$\therefore$ $\underline{x = 3.5}$.

Example 10 Find x.

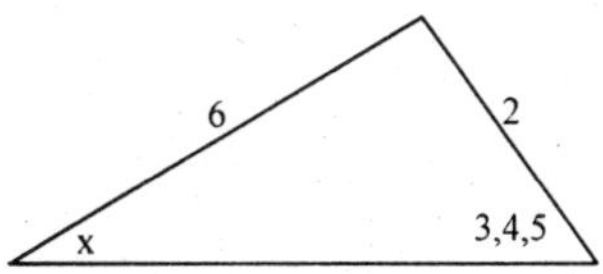

If x)q r s then $\frac{6r}{s}=\frac{2\times4}{5}$

$\therefore$ $\frac{r}{s}=\frac{4}{15}$ $\therefore$ $\underline{\text{x) - , 4, 15}}$

Example 11 Find angle A.

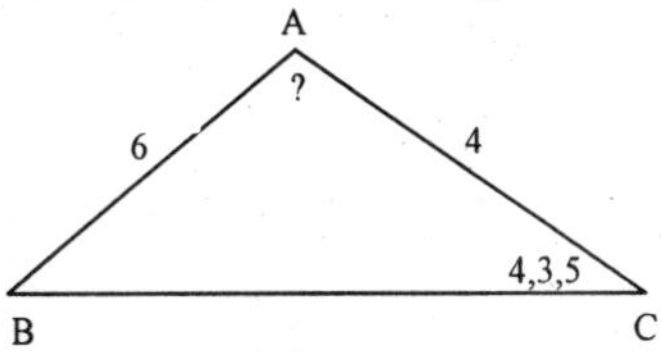

We can find B first and then A.

If B is given by B)Q, R, S then $\frac{6R}{S}=\frac{4\times3}{5}$.

$\therefore$ $\frac{R}{S}=\frac{2}{5}$ $\therefore$ B) -, 2, 5.

We then find 180° – (B+C):

B :	$\sqrt{21}$	2	5	
C :	4	3	5	+
B+C :	$4\sqrt{21}-6$	$8+3\sqrt{21}$	25	

$\therefore$ A = 180 – (B+C) = $\underline{6-4\sqrt{21} \quad 8+3\sqrt{21} \quad 25}$

Alternatively, since A is the sum of the angles complementary to B and C:

That is:

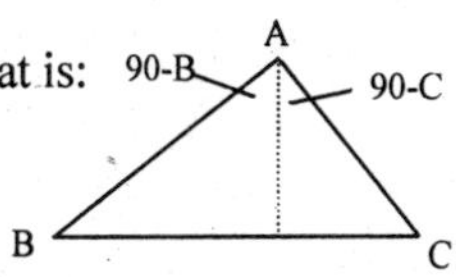

90–B :	2	$\sqrt{21}$	5	
90–C :	3	4	5	+
A :	$6-4\sqrt{21}$	$8+3\sqrt{21}$	25	

Example 12 Find AC.

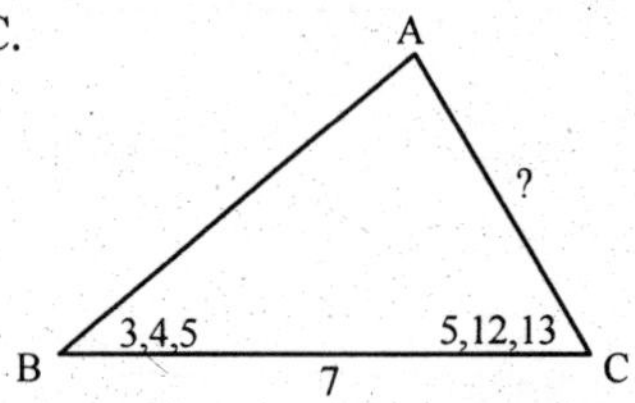

Similarly A is the sum of the triples complementary to B and C.

$$\therefore \begin{array}{rrrr} & 12 & 5 & 13 \\ & 4 & 3 & 5\ + \\ \hline A = & 33 & 56 & 65 \end{array}$$

Then, using the sine formula: $AC \times \frac{56}{65} = 7 \times \frac{4}{5}$

$\therefore\ AC = \frac{7 \times 4 \times 65}{5 \times 56} = \underline{6.5}$.

Exercise B Find x in the triangles below, giving the angles in triple form:

1.

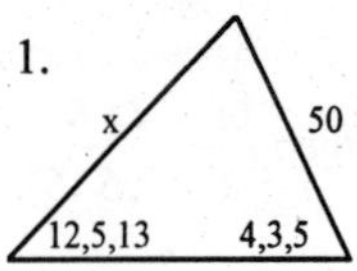

2.

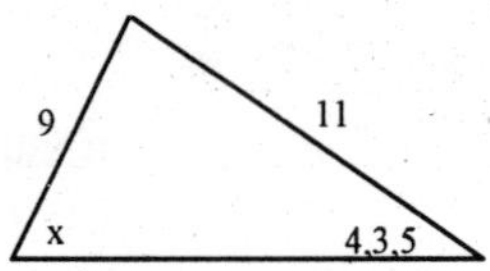

3.

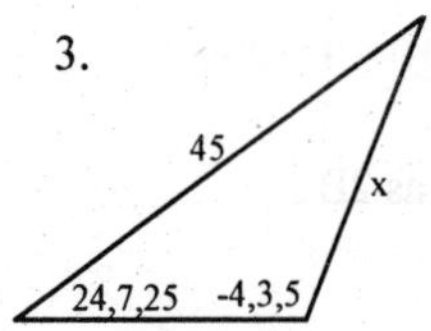

4.

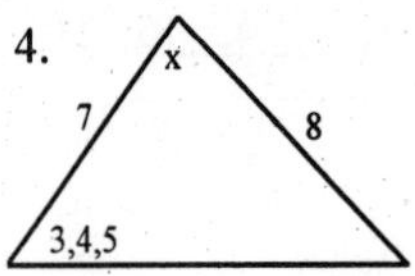

5.

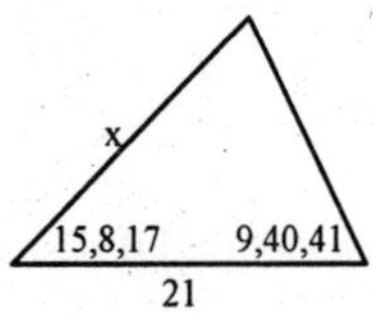

6.

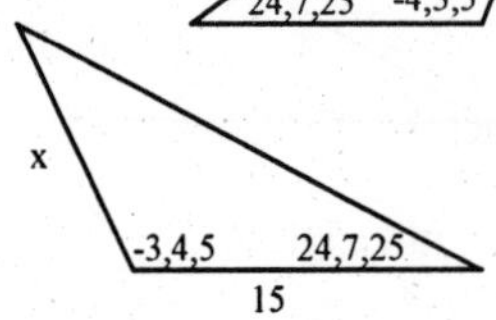

THE CODE-NUMBER FORMULA

For a triangle the formula is:

$$\frac{c(B-C)}{c(B+C)} = \frac{b+c}{b-c}$$

where c(B+C) and c(B–C) means the code numbers of the angles (B+C) and (B–C) respectively and b and c are the lengths of two sides of the triangle.
This comes under the Vedic Sutra *By Addition and By Subtraction.*
A proof is given on Page 162.

<u>Example 13</u> Find angles B and C.

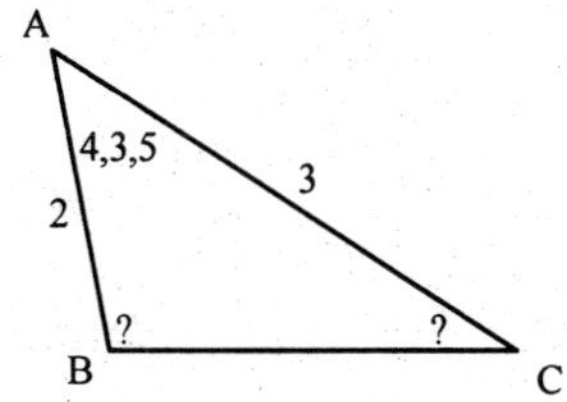

Given two sides and the included angle this formula gives the other two angles.

<u>Method (i)</u>: c(A) = 3 $\therefore$ c(B+C) = $\frac{1}{3}$ (as A and (B+C) are supplementary).

$\therefore$ using the code-number formula: $\frac{c(B-C)}{c(B+C)} = \frac{b+c}{b-c}$ c(B–C) = $\frac{1}{3} \times \frac{3+2}{3-2} = \frac{5}{3}$.

Adding the triples for B+C and B-C will give a triple for 2B, which we can then halve to get B:

B + C	–4	3	5	
B – C	8	15	17	+
2C	–77	–36	85	
C	–8	36	-	

Note that as 2B is a reflex angle we change the sign of the first two elements of B (see Page 16).

$\therefore$ <u>B) –2 9 -</u>

For C we subtract the triples for (B+C) and (B–C):

B + C	–4	3	5	
B – C	8	15	17	–
2C	13	84	85	
C	98	84	-	= <u>7 6 -</u>

<u>Method (ii)</u>: we can work entirely with the code numbers.
c(B+C) = $\frac{1}{3}$, c(B–C) = $\frac{5}{3}$ as before, but since a code number triple has half the angle of the triple, the angle in the triple 1, 3, - is ½(B+C) and the angle in the triple 5, 3, - is ½(B–C) so that adding these will give a triple for B:

½(B+C)	1	3	
½(B–C)	5	3	+
B	–4	18	= <u>–2 9</u>

$\therefore$ <u>B) –2 9 -</u>

Similarly for angle C:

½(B+C)	1	3	
½(B–C)	5	3	–
B	14	12	= <u>7 6</u>

$\therefore$ <u>C) 7 6 -</u>

<u>Example 14</u> Find angles B and C.

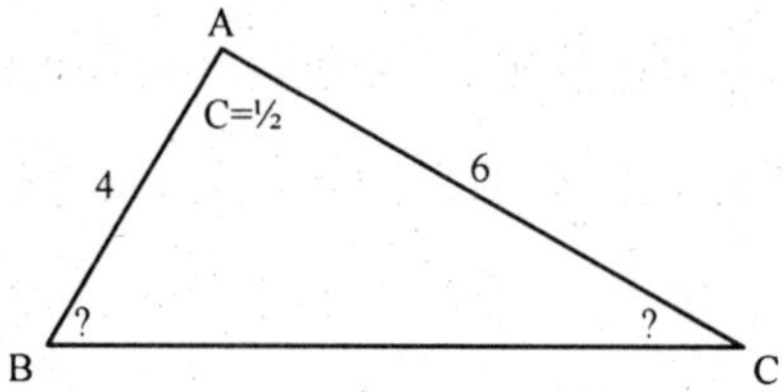

c(A) = 1, 2 ∴ c(B+C) = 2, 1

∴ using the formula: c(B–C) = $2\times\frac{10}{2}$ = 10.

(i)

B + C	3	4	5
B – C	99	20	101 +
2B	217	456	505
B	722	456	-
=	19	12	-

B + C	3	4	5
B – C	99	20	101 –
2C	377	336	505
=	882	336	-
C	21	8	-

(ii)

½(B+C)	2	1
½(B–C)	10	1 +
B	–19	12

½(B+C)	2	1
½(B–C)	10	1 –
C	21	8

Exercise C Find angles B and C:

1. 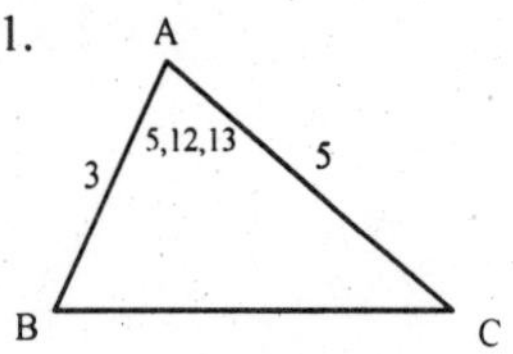

2. 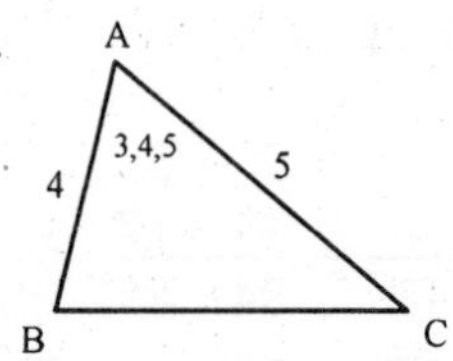

3. 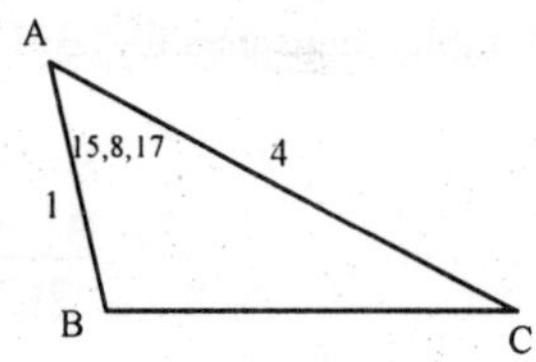

4. 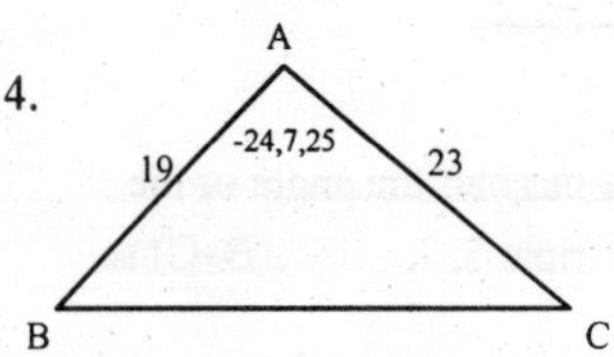

5. 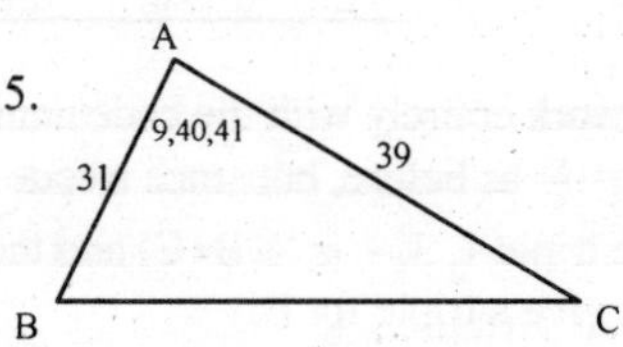

8 FURTHER APPLICATIONS OF TRIPLES

SOLUTION OF EQUATIONS

Certain special types of equation are related to triples. An outline of some interesting examples is given here.

Example 1 Solve $x + \frac{1}{x} = \frac{29}{10}$.

An equation of this type, which has rational solutions, is related to a perfect triple in which the product of the code numbers is the denominator on the right-hand side. If the terms on the left-hand side are added as they are here, the numerator on the right is the hypotenuse of the triple.
And if they are subtracted then the numerator is the base of the triple.

This can be seen from the equation: $\frac{a}{b} \pm \frac{b}{a} = \frac{a^2 \pm b^2}{ab}$

$\therefore$ since $10 = 5 \times 2$ and $5^2 + 2^2 = 29$, $\underline{x = \frac{5}{2} \text{ or } \frac{2}{5}}$.

Example 2 Solve $x - \frac{1}{x} = \frac{16}{15}$

Here $15 = 5 \times 3$ and $5^2 - 3^2 = 16$ $\therefore$ $\underline{x = \frac{5}{3} \text{ or } -\frac{3}{5}}$.

Example 3 If $\cos\theta + \sec\theta = \frac{58}{21}$ find $\cos\theta$

Since $21 = 7 \times 3$ and $7^2 + 3^2 = 58$ $\therefore$ $\underline{\cos\theta = \frac{3}{7}}$ and θ) 3, –, 7

Example 4 If $\cot\theta - \tan\theta = \frac{9}{20}$ find $\cot\theta$

$20 = 5 \times 4$ and $5^2 - 4^2 = 9$ ∴ $\cot\theta = \frac{5}{4}$ or $-\frac{4}{5}$

Example 5 If $e^x + e^{-x} = \frac{17}{4}$ find e^x

We get $e^x = 4$ or ¼.
Equations of this type arise when dealing with hyperbolic functions, see Chapter 11.

Example 6 Solve $3x^2 - 10x + 3 = 0$.

Equations with symmetrical coefficients, like this one, have roots which are reciprocal to one another. Here we see that $3 = 3 \times 1$ and $3^2 + 1^2 = 10$.
That is, the first or last coefficient can be split into two factors, the sum of the squares of which is the middle coefficient (disregarding its sign).
These factors give the answer: $x = \frac{1}{3}$ or $\frac{3}{1}$ (x cannot be negative).

Example 7 Solve $3x^2 + 10x + 3 = 0$.

This is just the same except that both answers are negative: $x = -\frac{1}{3}$ or -3 (x cannot be positive).

Example 8 Solve $6x^4 - 35x^3 + 62x^2 - 35x + 6 = 0$.

This equation also has symmetrical coefficients, the roots being in reciprocal pairs.
By first making the substitution $y = x + \frac{1}{x}$ we can break the quartic down into a quadratic in y: $6y^2 - 35y + 50 = 0$
in which the first two coefficients are the same as the first two (or the last two) of the quartic, and the last is the middle coefficient of the quartic reduced by the first and last coefficients, i.e. $62 - 6 - 6 = 50$.

∴ $(2y - 5)(3y - 10) = 0$
∴ $y = \frac{5}{2}$ or $\frac{10}{3}$.

So $x + \frac{1}{x} = \frac{5}{2}$ or $x + \frac{1}{x} = \frac{10}{3}$
∴ $x = 2$ or $\frac{1}{2}$ or 3 or $\frac{1}{3}$.

Example 9 Solve $6x^4 + 35x^3 + 62x^2 + 35x + 6 = 0$.

Similarly $y = x + \frac{1}{x}$ leads to $6y^2 + 35x + 50 = 0$.
$\therefore$ $(2y + 5)(3y + 10) = 0$ and $y = -\frac{5}{2}$ or $-\frac{10}{3}$ so that $x = -2, -\frac{1}{2}, -3, -\frac{1}{3}$.

COMPLEX NUMBERS

Multiplying complex numbers is similar to adding triples and dividing complex numbers is similar to subtracting triples.

Example 10 Find $(4 + 3i)(12 + 5i)$.

$$\begin{array}{ccc} 4 & + & 3i \\ 12 & + & 5i \\ \hline (4\times12-3\times5) & + & (3\times12+4\times5)i \end{array} = 33 + 56i.$$

That is, we multiply vertically and subtract, then multiply crosswise and add, as if we were getting the first two elements of the **sum** of two triples.

Example 11 Find $\dfrac{4+3i}{12+5i}$

$$\begin{array}{ccc} 4 & + & 3i \\ 12 & + & 5i \\ \hline \end{array}$$

$[(48 + 15) + (36 - 20)i] \div 13^2 = \dfrac{63+16i}{169}$ (as if we were getting the first two elements of the **difference** of two triples).

For division we must also divide by the square of the third element of the denominator triple. (That is, $12 + 5i \rightarrow 12, 5, 13$.)

This is far easier than the usual division method which involves multiplying top and bottom of the fraction by the complex conjugate of the bottom.

Proof: $\dfrac{a+bi}{c+di} = \dfrac{a+bi}{c+di} \times \dfrac{c-di}{c-di} = \dfrac{ac+bd+(bc-ad)i}{c^2+d^2}$ and the 3rd element of c+di is $\sqrt{c^2+d^2}$.

In the case of the square root of a complex number the difference between the conventional and triple methods is even more striking.

<u>Example 12</u> Find $\sqrt{3+4i}$

We know that when the square root of a complex number is taken the argument gets halved and the modulus gets "square-rooted". But, the angle in a code number triple is half the angle in the triple.

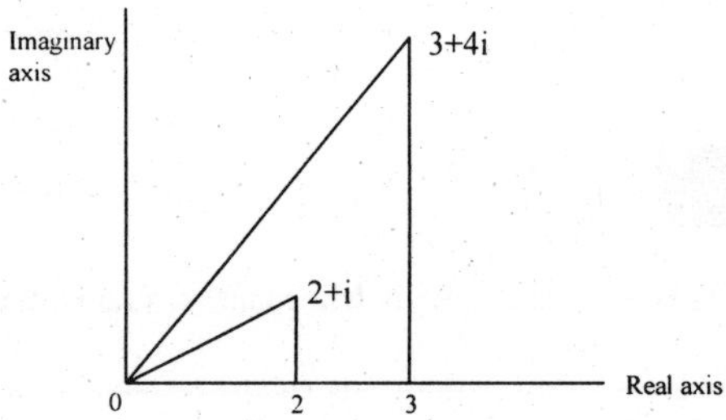

The complex number 3+4i relates to the triple 3,4,5 whose code numbers are 2,1.
These numbers are the coefficients of the answer: $\sqrt{3+4i} = \pm(2+i)$.

Since the modulus of 3+4i is 5 we require the modulus of the square root to be $\sqrt{5}$ and the modulus of 2+i is indeed $\sqrt{5}$.

<u>Example 13</u> Similarly, $\sqrt{8+6i} = \pm(3+i)$, as the code numbers of 8,6,10 are 3,1.

$$\sqrt{5+12i} = \pm(3+2i).$$
$$\sqrt{3-4i} = \pm(2-i).$$

<u>Example 14</u> Find $\sqrt{6+8i}$

The code numbers of 6,8,10 are 2,1 but here the modulus of 2+i is $\sqrt{5}$ and $\sqrt{10}$ is required.
We must therefore give the answer as $\pm\sqrt{2}(2+i)$.

<u>Example 15</u> Find $\sqrt{4+3i}$.

Similarly the code numbers of 4,3,5 are 3,1, but 3+i has a modulus of $\sqrt{10}$ when $\sqrt{5}$ is required: we must therefore divide by $\sqrt{2}$: $\sqrt{4+3i} = \pm\frac{3+i}{\sqrt{2}}$.

We simply take the code numbers of the triple, compare the moduli and add an appropriate factor if necessary.

CONICS

The conics are the various curves that are produced by cutting a cone at different angles and places. They include the parabola, ellipse, hyperbola and circle.

Parabola

Example 16 A parabola can also be defined as the path of a point whose distance from a fixed point (or focus) is always equal to its distance from a fixed line.
This means that if the fixed point is taken as the origin and the fixed line is the line x = –2, then the distance of a point on the parabola from the origin must equal its distance from the line, which is its x-coordinate plus the 2 units between the fixed line and the fixed point:

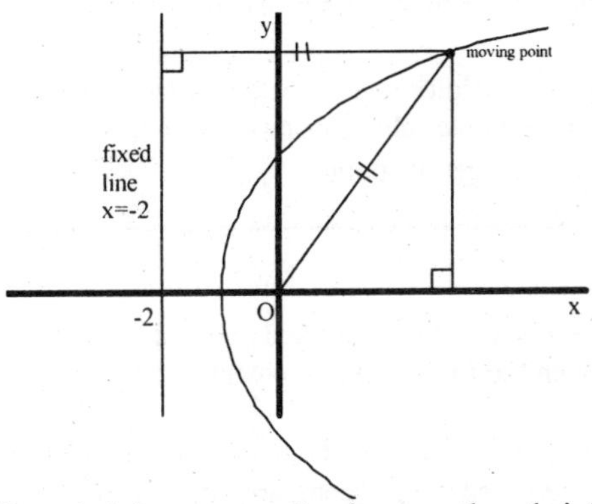

Thus those triples which have their hypotenuse 2 units more than their base, e.g. 3,4,5 can give points on this parabola:

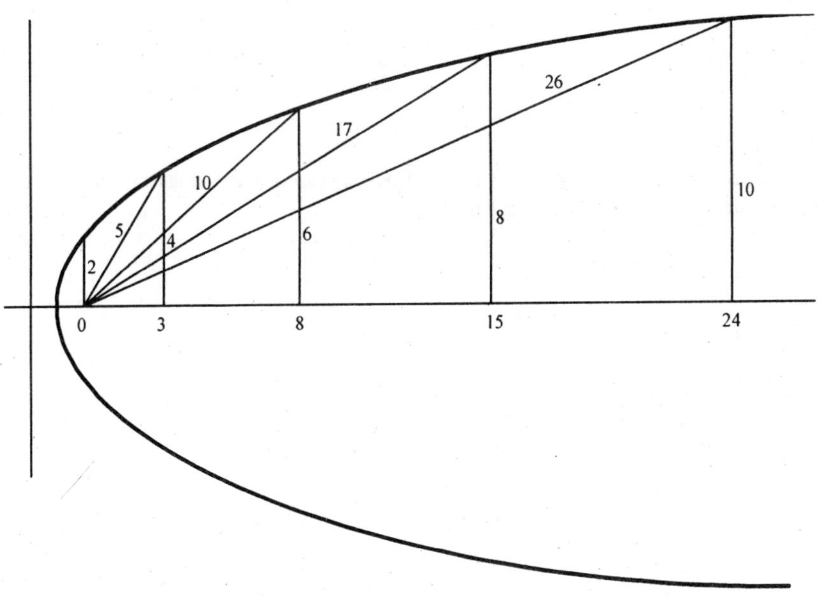

The 3,4,5 triple fits as shown, as does 15,8,17.

The 4,3,5 triple needs multiplying by 2 for the outer elements to differ by 2, and similarly for the 12,5,13 triple.

Thus any triple can be used by multiplying it through by an appropriate value.
Also 0,2,2 gives the point above O and the triple –1,0,1 gives the vertex.

The code numbers of the six triples used above are 0, 1, 2, 3, 4 and 5 and so we can obtain further triples inbetween these if required.

For example if a point is required between the triples with code numbers 2 and 3 we find the triple with code number $2\frac{1}{2}$ or 5, 2. This gives the triple 21, 20, 29 and since the outer elements here differ by 8 we divide it through by 4. This gives us the point $(\frac{21}{4},5)$.

For points to the left of the origin triples with a negative first element must be used.
For example the difference of the outer elements in –3, 4, 5 is 8 (as 5– –3=8). We therefore divide this triple through by 4 and get the point $(-\frac{3}{4},1)$.

In general the parabola $y^2 = 4ax$, in which the fixed line and fixed point are 2a apart, can be constructed by triples in which the first and last elements differ by 2a.

But points must be measured from the focus which is always a units from the vertex.
So if for example $y^2 = 2x$ is required we can compare this with $y^2 = 4ax$ to get $a = \frac{1}{2}$.
If 3,4,5 is then chosen to give a point on the parabola we must divide this through by 2 so that the first and last elements differ by 2a. This gives the point $(1\frac{1}{2},2)$.

All these points are measured from the focus so that if an origin is at the vertex of the parabola, as in $y^2 = 2x$, the value of a would always have to be added to the first coordinate.

Ellipses, hyperbolas and in fact many curves that can be expressed with the sum or difference of two squares (see later) can be plotted using triples.

Ellipse

Now consider the triple 2x, 3y, 12.
Being a triple it follows that $(2x)^2 + (3y)^2 = (12)^2$, or $4x^2 + 9y^2 = 144$.
Since this is the equation of an ellipse the triple **2x, 3y, 12 represents an ellipse.**

Example 17 Plot the ellipse 2x, 3y, 12.

Suppose first of all that we want to fit 4,3,5 into the ellipse.
That is, we want to find the values of x and y such that the triples 2x,3y,12 and 4,3,5 are equal.
Set out the triples:

2x	3y	12
4	3	5

then cross-multiply in the first and last columns to get $x = \dfrac{12\times 4}{2\times 5} = 4.8$.

And cross-multiply in the middle and last columns to get $y = \dfrac{12\times 3}{3\times 5} = 2.4$.

So the ellipse passes through the point (4.8,2.4). See the next diagram.

Exactly the same procedure with any other triple will yield another point on the ellipse. Any triple can be used: to get the semi-major axis, for example, use 1,0,1, which gives (6,0), or 0,1,1 which gives (0,4).

Note: 1. This method avoids the necessity of calculating square roots, which usually occur when choosing a value for x and then finding the corresponding value of y.

2. The ellipse and hyperbola shown on the next page were plotted by a computer using exactly the method described here. The program makes use of just 7 perfect triples.

3. Ellipses etc. whose centre is not at the origin can also be plotted (see Example 19).

4. The general ellipse is expressed by the triple **bx,ay,ab**, where a and b are the lengths of the semi-major and semi-minor axes. So the ellipse above can be expressed as 4x,6y,24. This means that the ratio of the coefficients in the first two elements is the ratio of the semi-minor to the semi-major axes: 2:3 in the example above.

5. If the last element of the ellipse triple is 1 we get the semi-major and semi-minor axes: from Example 17 we get $\frac{x}{6},\frac{y}{4}$,1 so the axes are 6 and 4 units respectively.

6. The relationship between a, b and the eccentricity, e, can also be expressed as a triple: ae,b,a (the reverse of this for the hyperbola).

7. This method can be extended to 3-dimensional conics (see later).

Example 18 Plot the hyperbola 4, 3y, 2x.

Note that in the case of the ellipse we had an x-element, a y-element and a number, and that this order is reversed for the hyperbola.

As before we put any chosen triple under the given one:

4	3y	2x
4	3	5

We cross-multiply in the first and last columns, starting in the top left corner this time:

$x = \frac{4\times5}{2\times4} = 2.5.$

Then in the first two columns starting in the same place: $y = \frac{4\times3}{3\times4} = 1.$

So we plot the point (2.5,1).

The triple 5,12,13 gives the point (5.2,3.2). And so on.

The curve is shown below.

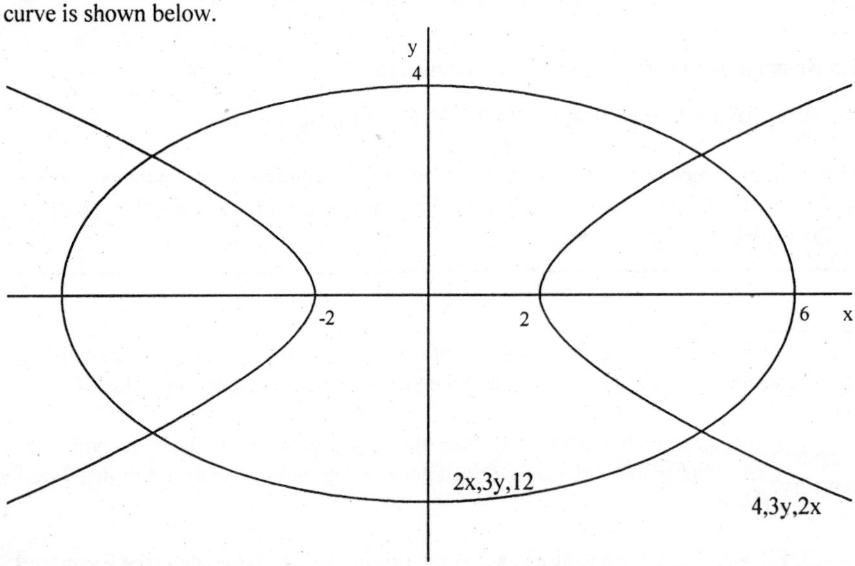

Example 19 Plot x–2, y+1, 3.

This represents a circle with its centre 2 units to the right of the origin and 1 unit down.

The procedure is the same as for the ellipse (a circle is a special kind of ellipse) except that 2 is added to the x value and 1 is subtracted from the y value:

x–2	y+1	3
4	3	5

$x = \frac{3\times4}{1\times5} + 2 = 4.4,\ y = \frac{3\times3}{1\times5} - 1 = 0.8.$ So (4.4, 0.8) lies on the circle.

OTHER TRIPLE-RELATED CURVES

As already mentioned other curves can be plotted if they can be expressed as a sum or difference of squares.

Example 20 Given the equation $x^2y^2 - 16y^2 + 1 = 0$ for example, a rearrangement gives: $x^2 + \frac{1}{y^2} = 16$.

Here we have a sum of squares on the left-hand side and so the equation can be written as a triple: $x, \frac{1}{y}, 4$.

We can fit triples to this in a similar way to the ellipse example earlier (the triple is similar to that for the ellipse in that it has an x term followed by a y term).

$$\begin{array}{ccc} x & \frac{1}{y} & 4 \\ 4 & 3 & 5 \end{array}$$

For 4,3,5 we get $x = \frac{4\times4}{1\times5} = 3.2$ and $\frac{1}{y} = \frac{4\times3}{1\times5}$, so $y = \frac{5}{12}$. Therefore $(3.2, \frac{5}{12})$ lies on the curve.

As before any triple can be used and all such curves are symmetrical about both axes.

Example 21 Similarly, given $x, xy, 4$ we get $\begin{array}{ccc} x, & xy, & 4 \\ 4 & 3 & 5 \end{array}$ and $x=\frac{16}{5}$ and $xy=\frac{12}{5}$ giving $y=\frac{3}{4}$.

Alternatively, we may consider $(x, \frac{1}{y})$, (see Example 20), as a point A, say, on a circle of radius 4.

Then consider the behaviour of points (x,y) as A goes around the circle from, say, (0,4).Thus any triple curve can be sketched by referring to the appropriate circle centred on the origin.

DIFFERENCE AND SUM OF TWO SQUARES

The formula, $a^2 - b^2$, $2ab$, $a^2 + b^2$ for generating triples from the code numbers a,b involves the quantities $a^2 - b^2$ and $a^2 + b^2$, and so the problem of expressing a given number as the difference or sum of two squares is related to triple generation.

A method for expressing an integer as the difference of two squares has been given by Sri Bharati Krsna Tirthaji in Chapter 31 of "Vedic Mathematics" (Reference 1). This method uses the identity: $\left(\frac{(a+b)}{2}\right)^2 - \left(\frac{(a-b)}{2}\right)^2 = ab$, which means that if a number can be expressed as a product ab it can be written as the difference of two squares. Since it is always possible to express an integer as a product of integers (and there will be an infinite number of ways of doing this if rational numbers are permitted) a solution is always possible.

<u>Example 22</u> Express 18 as a difference of two squares.

Using 18 = 18 × 1 we have a = 18, b = 1 and so

$$18 = \left(\frac{(18+1)}{2}\right)^2 - \left(\frac{(18-1)}{2}\right)^2 = (9\tfrac{1}{2})^2 - (8\tfrac{1}{2})^2.$$

Or, using 18 = 9 × 2 we have $18 = \left(\frac{(9+2)}{2}\right)^2 - \left(\frac{(9-2)}{2}\right)^2 = (5\tfrac{1}{2})^2 - (3\tfrac{1}{2})^2.$

<u>Example 23</u> Express 33 as a difference of two squares.

The number 33 occurs in the triple 33,56,65 whose code numbers are 7,4.
So $33 = 7^2 - 4^2$.
Any number which occurs as the first element of a triple whose code numbers are not both odd can be expressed as the difference of two squares in this way.

<u>Example 24</u> Express 82 and 41 as a sum of squares.

By Mere Observation $82 = 9^2 + 1^2$.
41 is half of 82 for which we used 9,1, so *By Addition and By Subtraction*:
½(9+1) = 5 and ½(9–1) = 4. Therefore $41 = 5^2 + 4^2$.

<u>Example 25</u> Express 37 and 74 as a sum of two squares.

Clearly $37 = 6^2 + 1^2$, and again *by addition and by subtraction* $74 = 7^2 + 5^2$.

<u>Example 26</u> Express 68 and 117 as a sum of squares.

By *proportion* in both cases: 68 = 4×17 and since $17 = 4^2 + 1^2$
$\therefore\ 68 = 8^2 + 2^2$ and $117 = 9\times13 = 9(2^2+3^2) = 6^2 + 9^2$.

<u>Example 27</u> Express 65 as a sum of two squares in two different ways.

$65 = 5 \times 13$ and $5 = 2^2+1^2$ and $13 = 3^2+2^2$.
So by triple addition and subtraction:

2	1		3	2	
3	2	+	2	1	–
4	7		8	1	

So $65 = 4^2 + 7^2 = 8^2 + 1^2$.

✵ Express 185 as a sum of two squares in two different ways.

Example 28 Express 29 as a sum of squares.

$29 \times 5 = 145$ and both 5 and 145 are easily expressible as a sum of squares: $5 = 2^2 + 1^2$, $145 = 12^2 + 1^2$.

So $2 - 12 = 25,10 = 5,2$. That is, $29 = 5^2 + 2^2$.

Alternatively, $29 \times 10 = 290$ and again both 10 and 290 are expressible as a sum of squares (3^2+1^2 and 17^2+1^2). So $17 + 3 = 50,20 = 5,2$.

Also as 29 occurs as the last element of the triple 21, 20, 29 the code numbers of this triple will give the answer.

INCIRCLES AND CIRCUMCIRCLES

The Incircle of a triangle is the circle which just touches all three sides.

The Circumcircle is the circle which passes through all the vertices.

It is interesting to find the radius of these circles for triples and relate them to the code numbers. The following are for triples in which the code number is a positive integer (so d=1).

The triple with code number c is $\mathbf{c^2 - 1,\ 2c,\ c^2 + 1.}$

Then the radius of the incircle, r, is given by $\mathbf{r = c - 1.}$

And the radius of the circumcircle, R, is given by $\mathbf{R = \frac{1}{2}(c^2 + 1).}$

Also the area, A, of the triangle is given by $\mathbf{A = (c - 1)c(c + 1).}$ That is, the product of three consecutive numbers the middle one of which is the code number.

So, for example, for c = 2, which generates the 3, 4, 5 triple, r = 1, R = 2½ and A = 6. When c = 3 the triple generated is 8, 6, 10 and this should not be divided through by 2 when applying the above formulae.

Below are the triples with code numbers 2, 3 and 4, drawn accurately, together with their incircles, circumcircles and the inner code number triangle whose base is the code number, whose height is 1 and whose hypotenuse bisects the angle in the triple.

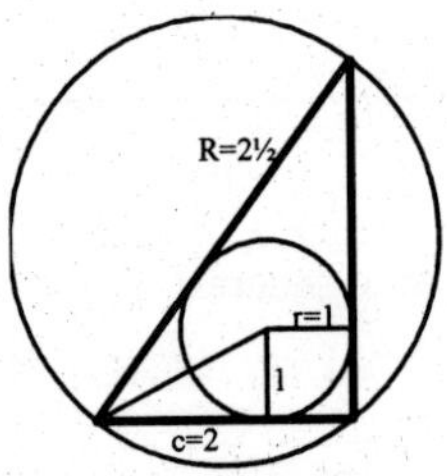

3, 4, 5

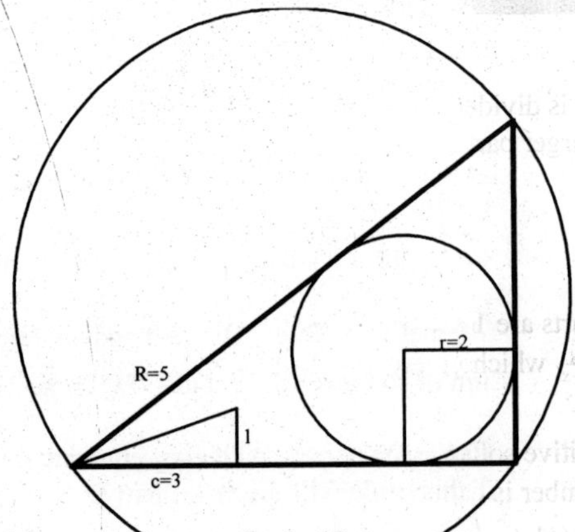

8, 6, 10

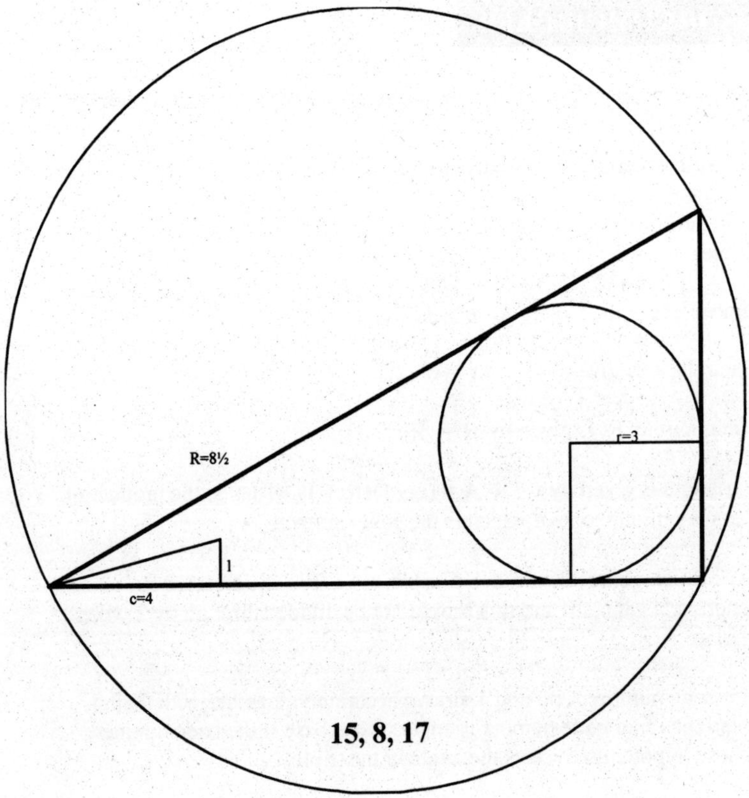

15, 8, 17

❋ Draw the triple, together with its inner code number triangle, for c=1.5 and $c = \sqrt{2}$.

THE GOLDEN TRIPLE

If a line is divided so that the ratio of the smaller part to the greater part is the same as the ratio of the larger part to the whole, line the line is said to be divided in the golden ratio.

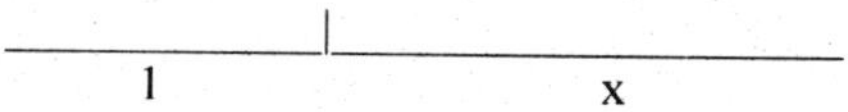

If the parts are 1 and x as shown then the above definition gives $1{:}x = x{:}1{+}x$ or $\frac{1}{x} = \frac{x}{1+x}$ which gives $1 + x = x^2$ on cross-multiplying.

The positive solution to the quadratic equation $x^2 - x - 1 = 0$ is $x = \frac{1}{2}(1 + \sqrt{5})$.
This number is called the golden number and is usually denoted by the Greek letter phi.
So $\phi = \frac{1}{2}(1 + \sqrt{5})$ and this is approximately 1.618.

In the three diagrams above we see that as c increases the inner code number triangle gets smaller in relation to the triple.
If you drew the triples with code numbers 1.5 and $\sqrt{2}$ you should have found that the "inner" code number triangle actually extends outside the triple.

This suggests that there exists a triple for which the vertical side of the code number triangle lies along the vertical side of the triple.

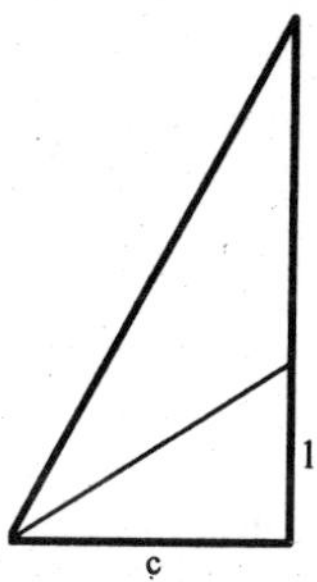

Since the base of the triple is $c^2 - 1$ and this is also c we get:

$$c^2 - 1 = c \quad \text{or} \quad c^2 - c - 1 = 0$$

This is the same equation as that obtained above for which the solution is the golden number.
So when $c = \phi$ the vertical sides of the triple and the code number triangle coincide.
This interesting result suggests that this particular triple be called the Golden Triple.

Since the height of this triple is 2c we see that this triple is half of a rectangle whose height is double the base.

This triple is also the code number triple for the triple –3, 4, 5: so that the golden triple is also directly related to the simplest of all the perfect triples.

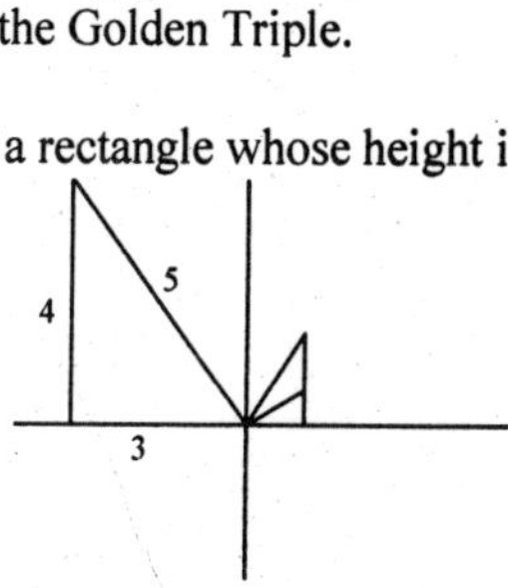

9 ANGLES IN PERFECT TRIPLES

Until now in this book angular answers have been left in triple form. In this chapter we will see how it is possible to find the angle in a given triple, in radians, and also how to find a triple containing an angle with a given number of radians.
It is not generally possible to find a triple containing an exact number of radians or degrees so we must content ourselves with answers to so many decimal places.
We will see that it is the code number of a triple that provides the link between the triple and its angle.

REVISION

It may be worth while looking again at some of the results and notation from Chapter 6.

1. Three numbers, e.g. 3, 4, 5 represent a triple,
   Two numbers, e.g. 3, 2 represent a pair of code numbers,
   One number, e.g. 3 represents a code number in which the second element is 1; i.e. 3 = 3, 1.

2. Finding the code numbers of a triple:
   e.g. for the triple 12, 5, 13 the code numbers, c,d = 12+13,5 = 25,5 = <u>5,1</u>.
   Generating a triple from code numbers:
   e.g. if c,d = 5,3 then the triple is $5^2 - 3^2$, $2 \times 5 \times 3$, $5^2 + 3^2$ = 16, 30, 34 = <u>8, 15, 17</u>.

3. We use a plus or a minus sign with a dot over it to denote triple addition and triple subtraction:

   e.g. 4, 3, 5 $\dot{+}$ 12, 5, 13 = 33, 56, 65 (using the *Vertically and Crosswise* addition method),

   $$3, 2 \ \dot{+}\ 5, 4 = 7, 22,$$

   $$4 \ \dot{+}\ 9 = 35, 13,$$

   $$2 \ \dot{+}\ 3 = 1 \qquad \text{(i.e. } 2,1 \ \dot{+}\ 3,1 = 1,1\text{)},$$

   $$1 \ \dot{-}\ 3 = 2, \quad \text{etc.}$$

4. We can write c(A) = 3 meaning that the code number for the triple with angle A is 3 etc.

Note a) $2 + 3 = 1$ states that the triple with code number 2 + the triple with code number 3 = the triple with code number 1; and so on.

Or to put it another way: the angle in the triple with code number 2 + the angle in the triple with code number 3 = the angle in the triple with code number 1.

b) The triples or code numbers can be shifted from one side of the equation to the other, just like ordinary equations:

e.g. if $2 + 3 = 1$ then $2 = 1 - 3$ (but $4 + 6 \neq 2$: you cannot multiply both sides).

c) For addition and subtraction of code numbers consisting of a single figure the usual addition and subtraction procedure simplifies:

For code numbers c and e we have

$$c + e = (ce - 1), (e + c) \qquad \text{e.g. } 4 + 9 = 35, 13,$$

$$c - e = (ce + 1), (e - c) \qquad \text{e.g. } 4 - 9 = 37, 5.$$

Exercise A Verify the following useful code number relations:

1. $4 + 13 = 3$
2. $3 - 4 = 13$
3. $2 - 3 = 7$
4. $1 - 2 = 3$
5. $5 - 7 = 18$
6. $4 + 8 = 3 + 21$
7. $3 - 5 = 8$
8. $2 - 12 = 3 + 17$
9. $3 - 23 = 4 + 30$
10. $3 - 4 = 15$
11. $11 + 50 = 9$
12. $2 + 2 - 7 = 1$

TRIPLES AND THEIR ANGLES

The table below is an extension of Table 1 from Chapter 6 (Page 52).
In addition to the triples and their code numbers the angle contained in each triple is shown, in radians, to 4 decimal places.

The first entry in the table is the triple for 90°, or $\frac{\pi}{2}$ radians.
The table can be used to confirm some of the relations listed in Exercise A.
For example we can see that $4 + 13 = 3$, since $0.4900 + 0.1535 = 0.6435$ radians.

It is important at this stage to appreciate that these two statements, one in code numbers and the other in angles, are equivalent.

TABLE 2

TRIPLE code no.	c		ANGLE A in rad.	2/c – A	TRIPLE code no.	c		ANGLE A in rad.	2/c – A
0	1	1	1.5708		255	16 ×2	257	0.1248	0.0002
3	2 ×2	5	0.9273		144	17	145	0.1175	0.0001
4	3	5	0.6435	0.0232	323	18 ×2	325	0.1110	0.0001
15	4 ×2	17	0.4900	0.0100	180	19	181	0.1052	0.0001
12	5	13	0.3948	0.0052	399	20 ×2	401	0.0999	0.0001
35	6 ×2	37	0.3303	0.0030	220	21	221	0.0952	0.0001
24	7	25	0.2838	0.0019	483	22 ×2	485	0.0908	0.0001
63	8 ×2	65	0.2487	0.0013	264	23	265	0.0869	0.0001
40	9	41	0.2213	0.0009	575	24 ×2	577	0.0833	0
99	10 ×2	101	0.1993	0.0007	312	25	313	0.0800	0
60	11	61	0.1813	0.0005	675	26 ×2	677	0.0769	0
143	12 ×2	145	0.1663	0.0004	364	27	365	0.0740	0
84	13	85	0.1535	0.0003	783	28 ×2	785	0.0714	0
195	14 ×2	197	0.1426	0.0002	420	29	421	0.0689	0
112	15	113	0.1331	0.0002	899	30 ×2	901	0.0666	0

The relationships in Exercise A (and there are many others) can be used in the construction of Table 2.
For example, if the angles for c = 1 and c = 2 are known then the angle for c = 3 is known since $1 \dot{-} 2 = 3$.
Similarly the angle for c = 7 is known since 2 – 3 = 7.

In fact it is possible to construct the table (extended indefinitely) starting only with the angles in the triples with code numbers 3, 4 and 5.

The angles in this table may also be calculated using series expansions as shown in reference 9, Chapters 9 and 10.

We may observe from this table that there is a simple relationship between the code numbers and their angles:

> the product cA is approximately equal to 2,

and gets closer to 2 further down the table, i.e. as c increases (see Page 96 for a derivation of $cA \approx 2$ for small A).

The difference between $\frac{2}{c}$ and A is given in the right-hand column.
This enables us to convert triples or code numbers to angles, or angles to code numbers or triples.

Example 1 Find the triple which contains an angle of 0.04 radians.

Since $c \approx \frac{2}{A}$, $c \approx \frac{2}{0.04} = 50$ which is the code number of the required triple.

So the triple is 2499, 100, 2501.

Example 2 Find the angle in the triple 1599, 80, 1601.

Since the code number is 40, the angle is $\frac{2}{40} = 0.05$ radians.

We discuss accuracy later (Page 163), but for now we may note that the angles calculated in this way are accurate:

to 2 decimal places for $c \geq 6$,
to 3 decimal places for $c \geq 13$,
to 4 decimal places for $c \geq 24$,
to 5 decimal places for $c \geq 52$.

Example 3 Find the angle in the triple 1512, 55, 1513.

$c = 55$ so $A = \frac{2}{55} = \frac{4}{110} = 0.03636$ to 5 decimal places.

If accuracy to 3 decimal places is sufficient then Table 2 becomes:

triples			angles
0	1	1	1.571
3	4	5	0.927
4	3	5	0.6435
15	8	17	0.490
12	5	13	0.395
35	12	37	0.330
24	7	25	0.284
63	16	65	0.249
40	9	41	0.221
99	20	101	0.199
60	11	61	0.181
143	24	145	0.166

further entries being unnecessary because $cA \approx 2$ will give the angles.

If only 2 decimal places are needed the table becomes:

0	1	1	1.57
3	4	5	0.93
4	3	5	0.64
15	8	17	0.49
12	5	13	0.395

There is also a simple device described in Reference 9, Chapter 9, for extending 2 decimal place accuracy to 4 decimal place accuracy.

Exercise B Find, to a suitable accuracy, the angles represented by the following triples and code numbers:

1. 840, 41, 841 2. c = 69 3. c = 51 4. 4899, 140, 4901 5. c = 20

Find code numbers for triples containing the following angles:

6. 0.025 7. 0.0444 8. 0.02 9. 0.008 10. 0.0143

In Questions 7 and 10 above it will be seen that there is a need to find a suitable fraction approximately equal to a given decimal. There are methods available in the Vedic system for doing this very efficiently and elegantly (the method given in Reference 1, Page 233, for example, but moving from left to right rather than right to left; see also the calculation of reciprocals in Reference 9, Chapter 1) but we do not go into this here. However the reader may have spotted that the fractions are $\frac{4}{90}$ and $\frac{1}{70}$ respectively.

FINDING THE ANGLE IN A GIVEN TRIPLE

So far we have dealt only with triples in which d=1, i.e. triples in Table 2.

Example 4 Find the angle in the triple 21, 20, 29.

The code numbers are 5,2 = 2½,1.
Since 2½ is between 2 and 3 the angle must be between 0.9273 and 0.6435rad.
We therefore use the c=2 triple, or the c=3 triple and find the difference between this and the given triple.

(i) 2½ − 3 = 8½,½ = 17,1 = 17. That is, 2½ = 3 + 17
Therefore using Table 2 the required angle is 0.6435+0.1175 = 0.7610 rad.

(ii) Alternatively, 2½ − 2 = 6,−½ = 12,−1 = −12 (this number is negative as 2 represents a larger angle than 2½)
So 2½ = 2 − 12 and therefore A = 0.9273 − 0.1663 = 0.7610 rad.

We will work with four decimal places in all approximate calculations from now on.

It will be seen that by consistently **subtracting the code number with the known angle** (c=2 and 3 above) **from the code number with the unknown angle** the sign of the result tells us whether the small angle is to be added to or subtracted from the known angle.

E.g. because 2½ – 2 = **–12** therefore A = 0.9273 – **0.1663.**

Example 5 Find the angle in the triple 55, 48, 73.

c,d = 128,48 = 8,3 and since this is close to 3: 8,3 – 3 = 27.

Therefore 8,3 = 3 + 27 and A = 0.6435 + 0.0740 = 0.7175 rad.

Example 6 Find the angle in the triple 56, 33, 65.

(i) The code number c = 11,3 is close to 3 so:

```
11   3
 3   1 –
36  –2  =  –18
```

So that 11,3 = 3 – 18 and A = 0.6435 – 0.1110 = 0.5325.

(ii) Since the code number is closer to 4 than 3:

```
11   3
 4   1 –
47   1
```

So 11,3 = 4 + 47.
The angle in the triple with code number 47 is not in the table.
However it can easily be found using $cA \approx 2$: for c=47, $A = \frac{2}{47}$ = 0.0425.
Therefore A = 0.4900 + 0.0425 = 0.5325.
And for c=47 we expect 4-figure accuracy.

(iii)

```
11   3
 2   1 –
25  –5  =  –5
```

therefore 11,3 = 2 - 5, so A = 0.9273 – 0.3948 = 0.5325.

We may be able to express code numbers as a sum or difference of code numbers with known angles as in (i) and (iii) above.
Alternatively, or if this is not possible, we select a known code number close to the given one (as in (ii) above) and use cA=2.

Note that divisions like $\frac{2}{47}$ etc. are easily carried out using the Vedic formula *By One More than the One Before* or *Vertically and Crosswise* (see Reference 1, Chapters 1, 26, 28 or Reference 8).

Exercise C Find the angles in the triples with code numbers:

1. 5,3 2. 11,2 3. 9,4 4. 11,5 5. 13,7 6. 16,5 7. 31,7 8. 15,7

9. 13,4 10. 34,7 11. 29,9 12. 43,10 13. 7,3 14. 10,3 15. 7,4

<u>Example 7</u> Convert the code numbers 23,9 to an angle.

The numbers are close to a ratio of 3:1 so:

$$\begin{array}{rr} 23 & 9 \\ \underline{3} & \underline{1} \; - \\ \underline{78} & \underline{4} = \underline{39,2} \end{array}$$

Therefore A = 0.6435 + $\frac{4}{39}$ = <u>0.746 to 3 d.p.</u>
We may only have 3 figure accuracy here as c = 19½ for the smaller angle.

If greater accuracy is required we could use the triple from Example 4:
0.7610)21, 20, 29 which has code numbers 5,2:

$$\begin{array}{rr} 23 & 9 \\ \underline{5} & \underline{2} \; - \\ \underline{133} & \underline{-1} \end{array}$$

This gives A = 0.7610 – $\frac{2}{133}$ = <u>0.7460</u>

The standard triples in Table 2 allow all possible angles to be covered by this method from 0 to $\frac{\pi}{4}$ radians to an accuracy of 4 decimal places, except for a gap between 0.72 and 0.78 radians. This gap is neatly covered by the 0.7610)21, 20, 29 triple as illustrated above.
There are many other useful code number relations which can be made use of in different ways.

For example:

1 ∸ 5 = 2 ∔ 8 = 3,2
2 ∸ 12 = 3 ∔ 17 = 5,2
3 ∸ 23 = 4 ∔ 30 = 7,2
2 ∸ 4 = 9,2
1 ∸ 7 = 4,3
1 ∸ 4 = 5,3
3 ∔ 12 = 7,3
1 ∸ 9 = 5,4
3 ∔ 5 = 7,4

For angles beyond $\frac{\pi}{4}$ we can use the complementary or supplementary triple, as the following examples show.

<u>Example 8</u> Find the angle in the triple for which c,d = 33,20.

These code numbers are close to a ratio of 3:2, and from the relations listed just above we see that 3,2 = 1 ∸ 5.
We therefore find out how far 33,20 differs from 3,2:

$$\begin{array}{rr} 33 & 20 \\ \underline{3} & \underline{2} \; - \\ \underline{139} & \underline{-6} \end{array}$$

So c(A) = 1 ∸ 5 ∸ 139,6,
and therefore A= 1.5708 – 0.3948 – $\frac{12}{139}$ = <u>1.0897</u>.

Alternatively, since the angle required is greater than $\frac{\pi}{4}$ we can find the angle in the complementary triple and take it from $\frac{\pi}{2}$:

We find the code numbers of the complementary triple using *By Addition and By Subtraction* (see Page 59):

The code numbers of the complementary triple are 53,13 which are close to a ratio of 4:1 so:

$$\begin{array}{rr} 53 & 13 \\ \underline{4} & \underline{1} \;- \\ \underline{225} & \underline{-1} \end{array}$$

Therefore $A = \frac{\pi}{2} - (0.4900 - \frac{2}{225}) = \underline{1.0897.}$

The diagram on Page 62 is useful for locating the sort of angle corresponding to a particular code number pair.

Example 9 c,d = 4,17.

Here the angle is obtuse and so we use the supplementary triple and subtract it from π.
Using the *Transpose and Apply* Sutra (see Page 60) the code numbers of the supplementary triple are 17,4:

$$\begin{array}{rr} 17 & 4 \\ \underline{4} & \underline{1} \;- \\ \underline{72} & \underline{-1} \end{array}$$

Therefore $A = \pi - (0.4900 - \frac{2}{72}) = \underline{2.6792.}$

Exercise D Convert the following code numbers into angles (to 4 d.p.):

1. 17,7 (use 5,2) 2. 25,16 (use 3,2 or the complementary triple) 3. 61,24

4. 31,19 5. 67,30 6. 37,12 7. 70,17 8. 103,16 9. 87,44

10. 91,3 11. 11,9 12. 23,22 13. 31,21 14. 3,16 (use the supplementary triple)

15. –4,19

For negative angles we can use the result from Page 63: c(–A) = –c(A). So if for example the code number is –3 the angle is –0.6435.

To summarise:

(i) If the code number is large we just divide it into 2 (as in Examples 2, 3 or Question 10 above);

(ii) otherwise we subtract from the code numbers other code numbers which have a similar ratio, and for which we know the angle, divide the result into 2, and add or subtract the two angles (as in Examples 6(ii) and 7-10).

(iii) For some triples there is a short cut available if the code numbers (or their complement or supplement) are the sum or difference of known code numbers (as in Example 6(iii)).

(iv) For angles greater than $\frac{\pi}{4}$ we use the complementary or supplementary triple (as in Examples 8,9).

FURTHER APPLICATIONS OF CODE NUMBERS

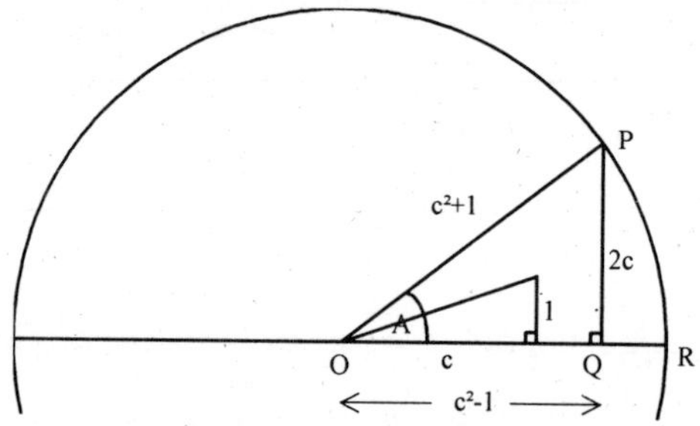

The diagram above shows the triple generated from the code number c.
The triple generated is $c^2-1, 2c, c^2+1$ and has the angle A which is double the angle in the code number triple.

> Example 10 Solve in the range $0 \le A \le 2\pi$ the equation $2\cos A - \sin A = 1$.
>
> If c is the code number of angle A then from the diagram above:
>
> $\sin A = \frac{2c}{c^2+1}$, $\cos A = \frac{c^2-1}{c^2+1}$, $\tan A = \frac{2c}{c^2-1}$
>
> So the equation becomes: $2(c^2-1) - 2c = c^2 + 1$.
> $\therefore\ c^2 - 2c - 3 = 0$
> $\therefore\ (c-3)(c+1) = 0$
> $\therefore\ c = 3$ or -1 and $A = 0.6435$ or $\frac{3\pi}{2}$ rad.

The type of equation solved above was solved in an earlier chapter where the solution was given in triple terms. These triples can now be translated into angles.
However the method of solution given above is quite different and uses code numbers rather than triples.

Derivation of $cA \approx 2$ for small A

Using the diagram above we can easily prove this result.
If A is measured in radians, the arc PR = $A(c^2 + 1)$.
$\therefore$ as $A \to 0$, $A(c^2 + 1) \to 2c$
$\therefore\ cA + \frac{A}{c} \to 2$
But as $A \to 0$, $c \to \infty$ $\therefore\ cA \to 2$.

> Example 11 Find the angle between the lines $3y = 4x - 12$ and $y = x + 10$.
>
> This is Example 6, Chapter 5 where the answer was found to be $7, 1, 5\sqrt{2}$.
> Now if 7,1 were code numbers we know that the angle in the triple with these code numbers is 0.2838 radians.
> The angle in the triple $7, 1, 5\sqrt{2}$ is therefore half of this, i.e. 0.1419 rad.

Example 12 Find the angle between $2y = 3x + 1$ and $3y = x + 7$.

By subtracting the line triples:

2	3	-	
3	1	-	–
9	7	-	

We therefore consider 9,7 as code numbers, find the corresponding angle and halve it:

9	7	
1	1	–
16	–2	∴ 9,7 = 1 – 8.

So the angle with code numbers 9,7 is 1.5708 – 0.2487 = 1.3221, and the angle between the lines is 0.6610 rad.

Example 13 Find the angle between $y = 4x$ and $3y = 7x$.

1	4			31	5	
3	7	–	and	6	1	–
31	5			191	–1	

So the angle as a triple is 31, 5, -
and this in radians is $\frac{1}{2}(0.3303 - \frac{2}{191}) = 0.1599$.

FINDING A TRIPLE WITH A GIVEN ANGLE

Example 14 Find a triple containing an angle of 0.597 radians.

This angle is not small enough to use $cA \approx 2$ but we can choose the nearest known triple and use $cA \approx 2$ on the difference.
The nearest basic triple to this is 0.6435)4, 3, 5 with c=3.

Since $0.597 - 0.6435 = -0.0465$ and $c = \frac{2}{A} = \frac{2}{-0.0465} = \frac{4000}{-93} = -43.01$,

in terms of code numbers we have $c(0.597) = 3 - 43 = 130,40 = 13,4$.

These code numbers generate the triple: 153, 104, 185.

The accuracy of this result is given by $a = \frac{2x}{c^2}$ (as shown on Page 163,4), where c is the secondary code number (43.01 in this example), and x is the quantity that is left off c (0.01 here).
So the accuracy in this case is given by $a = \frac{0.02}{43^2}$, which is negligible compared to the error in using 0.6435 (which is given to only 4 decimal places).
The above triple will therefore have an angle of 0.5970 rad.

Note that, as in the converse process, if we consistently **subtract the base angle** (0.6435 in the last example) **from the given angle**, the sign of the result indicates whether we add to or subtract from the base code number.

<u>Example 15</u> Find code numbers for a triple with an angle of 0.667 rad.

Using c(0.6435) = 3 again, the difference between these two angles is 0.0235, and $c(0.0235) = \frac{2}{0.0235} = \frac{4000}{47} = 85.11$.

So the given angle has been split up into two angles whose code numbers we know.

Therefore c(0.667) = 3 + 85 = 254,88 = <u>127,44</u>, which are the required code numbers.

For accuracy $a = \frac{2x}{c^2} = \frac{0.22}{85^2} \approx \frac{0.22}{7000} \approx 0.00003$.

<u>Example 16</u> Find c for 0.895 rad.

$0.895 - 0.9273 = -0.0323, \frac{2}{-0.0323} = -61.92 \approx -62$.

So c(0.895) = 2 – 62 = 125,60 = <u>25,12</u>.

And $a = \frac{0.16}{4000} = 0.00004$.

Exercise E Find code numbers for the following angles:

1. 0.62 rad.	2. 0.69	3. 0.525	4. 0.986	5. 0.434
6. 0.666	7. 0.741	8. 0.544	9. 0.527	10. 0.977
11. 0.883	12. 0.891	13. 0.6	14. 0.5	

A REFINEMENT

It may have been noticed in the last three examples that when the final code number has been found some cancelling has occurred in each case: by 10, 2 and 5 respectively.

It is easy to see that if two code numbers which are to be added or subtracted are both odd, then the result will be divisible by 2. This has happened in Example 15.

It can also be shown (see Page 162) that if c is the first (base) code number, then if $c^2 + 1$ divides into the second part of the final code number pair, it will also divide into the first part.

In the last example, where $c=2$, $c^2+1=5$ and we saw that 125,60 could be divided by 5 to give 25,12.
This cancelling does not usually occur however, but since the second code number is rounded off at some point we can arrange things so that the cancelling takes place, as the following examples show.

Example 17 Find code numbers for 0.65 rad.

$0.65 - 0.6435 = 0.0065$ and $\frac{2}{0.0065} = \frac{12000}{39} = 308$.

We find here that as the two angles are close together the second code number is rather large. But being large we may make a small change in it without losing, say, 4-figure accuracy. And since the c^2+1 value is 10 we change 308 so that the second part of the result is divisible by 10.

3 + 308 = - , 311 and if this 311 was reduced to 310 it would be a multiple of 10.

We therefore change 3 + 308 to 3 + 307 which gives 920,310 = 92,31.

Once the last part (31) is found the first part can be found by multiplying by the base code number (3), and subtracting 1: i.e. $31 \times 3 - 1 = 92$. (But add 1 if the code numbers are being subtracted)

Example 18 Find code numbers for 0.498 rad.

$0.498 - 0.49 = 0.008$ and $\frac{2}{0.008} = 250$.

We therefore want 4 + 250 and since the c^2+1 value is 17 here we change 250 so that 4 + 250 is a multiple of 17.

$15 \times 17 = 255$ so we can change 4 + 250 to 4 + 251.

We know we have arranged to have 15 17s, so 15 is the last part of the answer. And $15 \times 4 - 1 = 59$ gives the first part. So the code numbers are 59,15.

Example 19 Find code numbers for 0.527 rad.

$0.527 - 0.49 = 0.037$ and $\frac{2}{0.037} = 54.054$ so we want 4 + 54.054.

We could use 4 + 54 here, but the code numbers would be quite large. The c^2+1 value is 17 and we find that if we change 54.054 to 53.8 the last part of the answer becomes 57.8 which is a multiple of 17.

So 4 + 53.8 = 12.6,3.4 (where $3.4 = 57.8 \div 17$ and $12.6 = 3.4 \times 4 - 1$) = 63,17.

So this refinement is not only useful when the secondary code number is large but also has general application in finding code numbers of angles.

Example 20 Find code numbers for 0.7 rad.

$0.7 - 0.6435 = 0.0565$, $\frac{2}{0.0565} = 35.40$ so we need $3 + 35.40$; and $c^2 + 1 = 10$.

The last part of $3 + 35.40$ is 38.4 and if we change this to $38\frac{1}{3} = \frac{115}{3}$ we have a multiple of 5 on top (which is a factor of 10).
Therefore $3 + 38\frac{1}{3} = 105, \frac{115}{3} = \underline{63,23}$.

Exercise F

Convert the following angles to code numbers:

1. 0.635 2. 0.664 3. 0.934 4. 0.486 5. 0.11 6. 0.081

7. 0.324 8. 0.636 9. 0.422 10. 0.17 11. 0.474 12. 0.82

13. 0.744 14. 0.576 15. 0.526 16. 0.516 17. 0.188

18. Find exact code numbers for a) 30° b) 45° c) 60°

Example 21 Find c(1.235).

Since this is close to 90° we can subtract it from $\frac{\pi}{2}$ and work with the remainder:
$\frac{\pi}{2} - 1.235 = 0.3358$.
Then $0.3358 - 0.3303 = 0.0055$, $\frac{2}{0.0055} = 364$.

$\therefore$ c(0.3358) = 6 + 364 = 59,10.
$\therefore$ *By Addition and By Subtraction* c(1.235) = $\underline{69,49}$.

Example 22 Find c(2.828).

This angle is close to 180° and $\pi - 2.828 = 0.3136$.
$0.3136 - 0.3303 = -0.0167$ and $\frac{2}{-0.0167} = -120$.

$\therefore$ c(0.3136) = 6 − 120 ≈ 6 − 117 = 19,3.
$\therefore$ *By Transposition* c(2.828) = $\underline{3,19}$.

Example 23 Find c(-3).

The angle is just over 180° and $\pi - 3 = 0.1416$.
$0.1416 - 0.1426 = -0.001$ so c(0.1416) = 14 − 2000 = 141,10.
We want to add 180° to the angle this represents, and the code numbers for this are 0,1, therefore c(−3) = 141,10 + 0,1 = $\underline{-10,141}$.

All these calculations can also be carried out in degree measure rather than radians. If the basic table (Table 2) is given in degrees we need to use $cA \approx 114.59$ instead of $cA \approx 2$ (where $114.59 = \frac{360}{\pi}$ approximately).

10 SINE, COSINE, TANGENT AND INVERSES

In the previous chapter we saw how to find the angle in a perfect triple or to find a perfect triple having a given angle. These methods therefore give inverse sines, cosines and tangents and sine, cosine and tangent of angles in perfect triples; and they can also be used when the triples are not perfect. However these methods are best suited to perfect triples and can be rather cumbersome when applied to non-perfect triples.

In this chapter we consider the solution of triples of the form:

A) x	-	1		x) p	-	r
A) -	x	1		x) -	q	r
A) x	-	1		x) p	q	-

in which x is required in each case and the other letters represent known values.

SINE, COSINE AND TANGENT

These can all be dealt with together.

We must first introduce the idea of a "near triple": three numbers which do not quite form a triple.
This was considered earlier in the Angle-Deficiency formula of Chapter 7: 7, 11, 13 for example does not satisfy the requirement of a triple, that it obeys Pythagoras' rule, since $7^2 + 11^2 = 170$, and $13^2 = 169$,
although the requirement is nearly satisfied. So 7,11,13 can be called a "near triple".

A triple for an angle of 0° is: 0°)1 0 1,
But for triples with very small angles we know that $\sin A \approx \tan A \approx A$, and this can be expressed by:

A)1 A 1 for small A

For example 0.01)1, 0.01, 1 is approximately true, in radians. Call it a "small triple" as it has a small angle.

Example 1 Find sin, cos and tan of 0.5 radians.

0.5 radians is close to 0.49 radians for which we have a perfect triple: 15,8,17.
So using ordinary triple addition:

0.49	15	8	17		a perfect triple
0.01	1	0.01	1	+	a "small triple"
0.50	14.92	8.15	17		

from which sin, cos and tan of 0.5 rad can be read off, as a fraction, which can be converted to a decimal if required.

Example 2 Similarly for 0.48, 0.47, 0.46 rad.

0.49	15	8	17
0.01	1	0.01	1
0.48	15.08	7.85	17

0.49	15	8	17
0.02	1	0.02	1
0.47	15.16	7.7	17

0.49	15	8	17
0.03	1	0.03	1
0.46	15.24	7.55	17

Accuracy

Of course the closer we are to one of the standard triples the more accurate the results will be, and in general, for sine and cosine, we need A above to be below 0.01 rad. for 4 decimal place accuracy, i.e. $A \leq 0.01$ for 4 decimal place accuracy,
$0.01 < A \leq 0.03$ for 3 decimal place accuracy,
$0.03 < A \leq 0.1$ for 2 decimal place accuracy.

For tangent the method is more accurate: $A \leq 0.04$ rad. for 4 decimal place accuracy,
$0.04 < A \leq 0.09$ rad. for 3 decimal place accuracy.

So tangent is generally accurate to one more decimal place than sine and cosine.
This is discussed in more detail on Pages 164.

Example 3 Find a near triple for 0.67 rad.

0.6435	4	3	5		a perfect triple
0.0265	1	0.0265	1	+	a small triple
0.67	3.9205	3.106	5		a near triple

Since the small angle is 0.0265 sine and cosine will be accurate to 3 d.p. and tangent to 4 d.p.

Note that we cannot expect more than 4 decimal place accuracy whilst we are using the angles in Table 2 which are given to 4 d.p.
For angles not close to a standard triple accuracy may be improved by using intermediate perfect triples.

Example 4 Find a near triple for 0.75 rad.

This is close to the triple 0.761)21 20 29:

0.761	21	20	29
0.011	1	0.011	1 –
0.75	21.22	19.769	29

which gives answers accurate to 4 d.p.

Example 5 Find a near triple for 0.6 rad.

This angle is not near any of the standard triples. It lies between 0.6435 rad, with c = 3
and 0.49 rad, with c = 4.
Since 0.6 lies closer to 0.6435 than 0.49 we may select c=$3\frac{1}{3}$= 10,3 as a close triple.
We find the angle in this triple, if it is not known, as in the previous chapter: 0.5829 rad.
Then:

0.5829	91	60	109
0.0171	1	0.0171	1 –
0.6	89.97	61.556	109

For angles between $\frac{\pi}{4}$ and $\frac{\pi}{2}$ radians, unless the angle is close to a known triple, it is best to find a triple for the complementary angle. Then for cosine read off the sine of the complementary triple, for sine read off the cosine and for tangent read off the cotangent.

Exercise A Find near triples for the following angles, stating the accuracy achieved.

1. 0.4 2. 0.31 3. 0.65 4. 0.73 5. 0.56

INVERSE COSINE AND INVERSE SINE

First note that from examining successive triples, after the first, in Table 2, Page 90, the ratio of the third element to the second element is approximately:

1, 1½, 2, 2½, 3, 3½ etc.

Example 6 Find x in the triple x) - 9 23.

Since 23:9 ≈ 2½:1 we select 0.3948)12 5 13 as a close triple:

x	-	9	23
0.3948	12	5	13

So the answer is close to 0.3948, the difference being given approximately by a fraction whose numerator is given by the determinant $\begin{vmatrix} 9 & 23 \\ 5 & 13 \end{vmatrix}$ and whose denominator is the product 12×23.

So $x = 0.3948 + \dfrac{\begin{vmatrix} 9 & 23 \\ 5 & 13 \end{vmatrix}}{12 \times 23} = 0.3948 + \dfrac{9\times13-23\times5}{12\times23} = 0.3948 + \dfrac{1}{138} = \underline{0.4020}$ to 4 d.p.

In general, for

x)	-	m	n
A)	p	q	r

$$x = A + \frac{\begin{vmatrix} m & n \\ q & r \end{vmatrix}}{pn} \qquad (1)$$

A derivation will be found on Page 164.

Also, for

x)	l	-	n
A)	p	q	r

$$x = A - \frac{\begin{vmatrix} l & n \\ p & r \end{vmatrix}}{qn} \qquad (2)$$

Accuracy is the same as that given earlier for sine and cosine:

4-figure accuracy if the fraction ($\frac{1}{138}$ above) is less than or equal to 0.01, i.e. $f \leq 0.01$,

3-figure if $0.01 < f \leq 0.03$,

and 2-figure if $0.03 < f \leq 0.1$.

In fact, as with sin, cos and tan the accuracy is also dependent on the angle A: the above limits are the most extreme values (see Page 165).

Of course we have found not only inverse sine $\frac{9}{23}$, but also inverse cosine $\frac{9}{23}$ (by subtracting the above answer from $\frac{\pi}{2}$).

Example 7 Solve x) - 11 24

24:11 ≈ 2:1 so we can use 0.49)15 8 17:

x	-	11	24
0.49	15	8	17

and $x = 0.49 - \frac{5}{360} = \underline{0.4761}$ which is actually accurate to 4 d.p.

Example 8 Solve x) 6 - 13.

Here we can solve the complementary triple y) - 6 13 and take the result from $\frac{\pi}{2}$.

y	-	6	13
0.49	15	8	17

$y = 0.49 - \frac{2}{195} = 0.4797$, so $x = 1.5708 - 0.4797 = \underline{1.0911}$.

Alternatively we can use formula (2) above.

Example 9 Solve x)1 - 3

y	-	1	3
0.3303	35	12	37

So $x = 1.5708 - 0.3303 - \frac{1}{105} = \underline{1.2310.}$

Or, using formula (2) above we again get: $x = 1.5708 - 0.3303 - \frac{1}{105} = \underline{1.2310.}$

Example 10 Solve x) - 23 25

It is not so clear here which is the best perfect triple to use because 23 and 25 are close. But if we **divide the sum of the numbers by their difference** we get 24 (48÷2) and we can look for a standard perfect triple which gives a value close to 24 when *By Addition and By Subtraction* is similarly applied to it.

The triple 12,5,13 gives a value of 25 when we divide the sum of the outer numbers by their difference and this is the closest we will get without using intermediate triples.

Using formula (2) above:

x	23	-	25
0.3948	12	5	13

So $x = 0.3948 - \frac{23\times13-25\times12}{5\times25} = 0.3948 + \frac{1}{125} = \underline{0.4028}$.

Similarly, to solve x) - , 23, 25 we would carry out the above calculation and take the result from $\frac{\pi}{2}$.

Example 11 Solve x)10 - 11.

The Addition and Subtraction number here is (10+11)÷(11–10) = **21**.
The 15,8,17 and 12,5,13 triples give **16** and **25** respectively.
This suggests we use a triple between 15,8,17 and 12,5,13 and so we choose the triple with code number c=4½: 77,36,85.

If the angle in the triple 77,36,85 is not known it can be found as shown earlier: the angle is 0.4373.

x	10	-	11
0.4373	77	36	85

So $x = 0.4373 - \frac{3}{396} = \underline{0.4297}$ to 4 D.P.

INVERSE TANGENT

This is the easiest case, and in fact it has already been dealt with in Examples 11, 12, 13 of the previous chapter.

Example 12 Solve x) 3 1 -

We know from Table 2 that the triple with code numbers 3,1 has an angle of 0.6435 radians.

So the angle in the x) 3 1 - triple must be $\frac{1}{2}$ of 0.6435 = 0.32175 rad.

So we can easily find angles in triples of the form x) n 1 -, where n is a positive integer, by halving the angle in the triple with code number n.

Example 13 Solve x) 1 3 -

The angle here is complementary to the one found in Example 12.
So $x = \frac{\pi}{2} - \frac{1}{2}(0.6435) =$ 1.2490 rad.

This also generalises for all triples of the form x) 1 n -.

Examples 14 and 15 Solve y) 3 –1 --
and z) 1 –3 --

These triples clearly have angles which are minus the angles found in Examples 12 and 13.
So y = –0.32175, and z = –1.2490 rad.

There are of course other answers, found by adding or subtracting multiples of 2π, but we will just find a single solution.

Example 16 Find x in x) 13 7 -

These numbers are close to a ratio of 2:1 and the angle for c=2 is 0.9273, so we find the difference between the code numbers 13,7 and 2,1:

13 7
2 1 –
33 1

The angle corresponding to a code number of 33 is $\frac{2}{33}$ (using $cA \approx 2$) or $\frac{6}{99}$,
so $2x = 0.9273 + 0.0606$, and $\underline{x = 0.49395}$.

This example illustrates how to deal with a triple in which the previous special cases do not apply. (Although the complementary triple could be used here.)

Accuracy here is as given on Page 91: 4 decimal places for $c \geq 24$ in this case.
Actually the difference between this answer and the correct one is 0.00001 but due to the rounding process there is a difference of 1 in the 4th decimal place when 4 d.p. is taken.

Example 17 x) 21 8 -

Here we see a ratio of approximately 3:1.

Using the 4,3,5 triple:

21	8
3	1 –
71	3

Therefore $x = \frac{1}{2}(0.6435 + \frac{6}{71}) = \underline{0.3640}$ to 4 d.p.

Example 18 Find x) 5 3 -

Instead of seeking the angle in the triple with code numbers 5,3 there is a useful result (see Page 94) we can use here: 1 – 4 = 5,3.

So the angle in the triple with code numbers 5,3 is 1.5708 – 0.4900 = 1.0808.

Therefore $x = \frac{1}{2}(1.0808) = \underline{0.5404 \text{ rad.}}$

Alternative method If the result 1 – 4 = 5,3 was not known, the following method may be used:

Treating 5,3 as code numbers the code numbers of the complementary triple will be 5+3,5–3 = 8,2 = 4,1 (see Page 59), and the angle in the triple with code number 4 is known to be 0.4900 rad.

So the angle in the triple 4, 1, - is $\frac{1}{2}(0.4900) = 0.2450$.

But since the angles in the complementary triples add up to $\frac{\pi}{2}$ the angles in the code number triangles must add up to $\frac{\pi}{4}$. So $x = \frac{\pi}{4} - 0.2450 = 0.5405$ rad.

In summary $x = \frac{\pi}{4} - \frac{1}{2}(0.4900) = \underline{0.5404 \text{ rad}}$.

This alternative method is always useful where the complementary triple is easier to work with than the given triple.

<u>Example 19</u> x) 9 4 -

$$\begin{array}{cc} 9 & 4 \\ \underline{2} & \underline{1} \\ 22 & -1 \end{array} -$$

$x = \frac{1}{2}(0.9273 - \frac{1}{11}) = \underline{0.4182}$ (note the minuses here).

However use of c=22 means that the last figure may be out by 1 unit in the last place (in fact it is not).
We could consider a known ratio closer to 9,4 and since the angle in the triple with code numbers 5,2 is 0.7610 we may use this:

$$\begin{array}{cc} 9 & 4 \\ \underline{5} & \underline{2} \\ 53 & 2 \end{array} -$$

So $x = \frac{1}{2}(0.7610 + \frac{4}{53}) = \underline{0.4182}$.

<u>Example 20</u> x) 3 17 -

Here we may solve y) 17 3 - and take the result from $\frac{\pi}{2}$:

$$\begin{array}{cc} 17 & 3 \\ \underline{6} & \underline{1} \\ 105 & 1 \end{array} -$$

So $x = \frac{\pi}{2} - \frac{1}{2}(0.3303 + \frac{4}{210}) = \underline{1.3961}$

<u>Example 21</u> Find $\tan^{-1}0.345$

$$\begin{array}{cc} 1 & 0.345 \\ \underline{3} & \underline{1} \\ 3.345 & 0.035 \end{array} - \quad = 669 \quad 7$$

Therefore $\tan^{-1}0.345 = \frac{1}{2}(0.6435 + \frac{14}{669}) = \underline{0.3322}$.

Exercise B Find x, giving answers to an appropriate accuracy:

1. x) - , 1, 3	2. x) - , 4, 7	3. x) 7, - , 15	4. x) - , 7, 18
5. x) - , 23, 24	6. x) - , 18, 19	7. x) 4, 1, -	8. x) 1, 4, -
9. x) 3, 1, -	10. x) 3, 2, -	11. x) 13, 4, -	12. x) 7, 3, -

11 HYPERBOLIC FUNCTIONS

We know that the triple **t) x y 1** represents simultaneously the equation $x^2 + y^2 = 1$ and the parametric pair x = cost, y = sint. The equation $x^2 + y^2 = 1$ represents a circle.

The equation $x^2 - y^2 = 1$ (which represents a hyperbola) can also be given in triple form as **t) 1 y x**, the reverse of the triple above. The parametric form is x = cosht, y = sinht (cosht is pronounced "cosh t" and sinht is usually pronounced "shine t"). Cosht and sinht are defined as

$$\cosh t = \tfrac{1}{2}\left(e^{t} + e^{-t}\right), \ \sinh t = \tfrac{1}{2}\left(e^{t} - e^{-t}\right) \qquad (1)$$

e here is the exponential number (approximately 2.7183). It is easy to show from these definitions that $\cosh^2 t - \sinh^2 t = 1$. These definitions lead to the study of the so-called Hyperbolic Functions which we will now look at briefly in terms of triples.

ADDITION AND SUBTRACTION

So just as **A) cosA sinA 1,** or **A) x y 1,** represents a circular-based triple,

A) 1 sinhA coshA or **A) 1 y x** represents a hyperbolic-based triple,

though "A" here does not represent a real angle (see Reference 11, Appendix 7).
It also follows from the above definitions that

$$\cosh(A+B) = \cosh A\cosh B + \sinh A\sinh B$$

$$\text{and} \quad \sinh(A+B) = \sinh A\cosh B + \cosh A\sinh B. \qquad (2)$$

Abbreviating coshA, sinhA, coshB, sinhB with cA, sA, cB, sB respectively we can write these results in triple form:

A)	**1**	**sA**	**cA**	
B)	**1**	**sB**	**cB**	+
A+B)	**1**	**sAcB+cAsB**	**cAcB+sAsB**	

To take a specific example: if $sA=\frac{3}{4}$ $cA=\frac{5}{4}$ $sB=\frac{5}{12}$ $cB=\frac{5}{12}$

then we have:

$$\begin{array}{c|ccc}
A & 1 & \frac{3}{4} & \frac{5}{4} \\
B & 1 & \frac{5}{12} & \frac{13}{12} \quad + \\
\hline
A+B & 1 & \frac{39+25}{48} & \frac{65+15}{48}
\end{array} \quad = \underline{48,\ 64,\ 80}$$

Avoiding the fractions completely we can write

$$\begin{array}{rcccc}
A & : & 4 & 3 & 5 \\
B & : & 12 & 5 & 13 \quad + \\
\hline
A+B & : & 48 & 64 & 80
\end{array}$$

in which we can see again that the *Vertically and Crosswise* pattern will give us the elements of the triple for A+B by

a) giving us 80 by multiplying vertically in the last two columns and adding,
b) giving 64 by cross-multiplying in the last two columns and adding,
c) giving 48 by multiplying in the first column.

For subtracting these hyperbolic triples the method is the same as for addition except that we subtract instead of adding:

$$\begin{array}{rcccc}
A & : & 4 & 3 & 5 \\
B & : & 12 & 5 & 13 \quad - \\
\hline
A-B & : & 48 & 14 & 50
\end{array} \quad \text{where} \quad \begin{array}{l} 50 = 65-15, \\ 14 = 39-25, \\ 48 = 4\times 12. \end{array}$$

Note that in cross-multiplying here to get the middle element we begin in the middle of the top triple (as in the similar case with the circular-based triples) i.e. the 3. We see also that these calculations are easier than the corresponding ones in Chapter 2 because the pairs of products are added in both cases for A+B and subtracted in both cases for A–B.

DOUBLE ANGLE

By putting A = B in equations (2) it can be seen that if A) 1 y x
then 2A) 1 2xy x^2+y^2

Or, for triples of the form **A) z y x**
then **2A) z^2 2xy x^2+y^2**

Example 1 Given $\sinh A = \frac{3}{4}$ find $\cosh 2A$

A) 4 3 5
2A) 16 30 34 = 8 15 17 so $\cosh 2A = \frac{17}{8}$.

Example 2 Show that $\tanh^2 A = \frac{\cosh 2A - 1}{\cosh 2A + 1}$ (Note $\tanh A = \frac{\sinh A}{\cosh A} = \frac{y}{x}$)

Since if A)1 y x then 2A)1 2xy x^2+y^2 the identity above can be written as $\frac{y^2}{x^2} = \frac{x^2+y^2-1}{x^2+y^2+1}$ but since $x^2 = y^2+1$ the right-hand side of this equation becomes $\frac{2y^2}{2x^2}$ which is clearly equal to the left-hand side.

EQUATIONS

Example 3 Solve $\cosh x = \frac{5}{3}$

We can write the solution in triple form as x)3 ±4 5.

From the original definitions (1) we can write the given equation as
$e^x + e^{-x} = \frac{10}{3}$.
The solution (for e^x) of equations of this type is considered on Page 76:
By Mere Observation we can see that since $\frac{10}{3}$ is the sum of two reciprocals
$e^x = 3$ or $\frac{1}{3}$,
and so $x = \pm \ln 3$.
(the evaluation of natural logarithms is dealt with in Reference 9, Chapter 7)

It is worth noting here that from the definitions (1) coshx cannot be negative, because e^x cannot be negative. Therefore in our expression of hyperbolic triples the outer elements should always be positive, though the middle element could be positive or negative.

Example 4 Solve $\sinh x = \frac{4}{3}$
Similarly $e^x - e^{-x} = \frac{8}{3}$ so $e^x = 3$, ($e^x = -\frac{1}{3}$ is not valid as e^x is always positive) and $x = \ln 3$.

Consider an alternative method: if A) c b a and $\cosh A = \frac{a}{c}$
then $e^A + e^{-A} = \frac{2a}{c}$, so $ce^{2A} - 2ae^A + c = 0$
and $e^A = \frac{2a \pm \sqrt{4a^2 - 4c^2}}{2c} = \frac{a+b}{c}$ since $a^2 = b^2 + c^2$.
So, given the triple A)3 4 5 we can immediately put $e^A = \frac{5+4}{3} = 3$.

Example 5 Solve $\cosh A = \frac{25}{7}$

The triple is A)7 ±24 25, so $e^A = \frac{25+24}{7} = 7$ or $\frac{1}{7}$.

Example 6 Solve $\cosh A = 3$

A)1 $\pm\sqrt{8}$ 3 gives $e^A = 3 \pm \sqrt{8}$.

Example 7 Solve $\tanh A = \frac{4}{5}$

A)3 4 5 gives $e^A = \frac{5\pm 4}{3} = 3$ or $\frac{1}{3}$.

Example 8 Solve $6\cosh A + 10\sinh A = 15$

The solution to this type of equation follows similar lines to that of the corresponding equation in circular functions as solved in Chapter 3.

Writing the equation as $6x + 10y = 15$:

A	1	y	x		
B	(8)	6	10	+	6 and 10 are the coefficients
A+B	8	15	17		8 is the vertical product, 15 is the RHS
B	4	3	5	-	this is B cancelled down
A	32	24	40		= 4, 3, 5

Example 9 Solve $3\cosh A + 5\sinh A = 6$

A	1	y	x	
B	(4)	3	5	+
A+B	4	6	$(\sqrt{52})$	
B	4	3	5	−
A	16	$30-3\sqrt{52}$	$5\sqrt{52}-18$	

These equations have one solution only.

Example 10 Solve 3coshA + 3sinhA = 5

Here we find the method appears to fail:

A	1	y	x	
B	0	3	3	+
A+B	0	5	5	
B	0	3	3	
A	0	0	0	!

However, if we write the equation as $5 - 3y = 3x$ we can set up the following chart in which the addition and subtraction are those used for circular functions:

A	1	y	x		
B	5	3	$\sqrt{34}$	+	
A+B	3x	(5x)	x$\sqrt{34}$		the x cancels out here
B	5	3	$\sqrt{34}$	-	
A	30	16	34		= 15, 8, 17

This method is valid because +B and –B are inverse operations.

12 APPLIED MATHEMATICS APPLICATIONS

SIMPLE HARMONIC MOTION

The equation **x = sint**, and its successive derivatives, can be represented by triples:

$$x = \sin t: \qquad t)\ -,\ x,\ 1$$

$$\dot{x} = \cos t: \qquad t)\ \dot{x},\ -,\ 1$$

$$\ddot{x} = -\sin t: \qquad t)\ -,\ -\ddot{x},\ 1$$

$$\dddot{x} = -\cos t: \qquad t)\ -\dddot{x},\ -,\ 1$$

etc.

Note that the x moves to and fro between the first and second positions of the triple, and that the sign changes when it moves to the right.

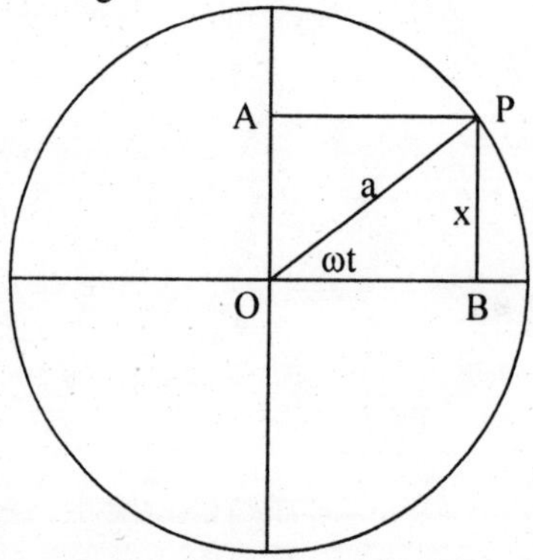

Suppose a point P moves with constant speed around a circle of radius a, as shown above.
PA and PB are perpendiculars from P onto the vertical and horizontal diameters.
The motion of A and B is called Simple Harmonic Motion.

If x is the height of P measured upwards from O and the angle is measured anticlockwise from OB then we can see that **x = asinωt.** This describes the motion of A.

This equation and its first two derivatives give:

$$\begin{array}{r|lll} & -, & x, & a \\ \omega t & \dot{x}, & -, & a\omega \\ & -, & -\ddot{x}, & a\omega^2 \end{array} \qquad \begin{array}{c}(1)\\(2)\\(3)\end{array}$$

Using these three triples it is an easy matter to solve problems in Simple Harmonic Motion by simply finding missing elements in the chart. We do not need the complex formulae usually associated with this topic: though these can be easily obtained from the chart (see later).

There are various important and useful observations that can be made from the above chart and these will be illustrated in the following examples. But note here that these three triples are equal as the angle is the same in each (ωt), and that the third element of each triple shows that the second triple is ω times the first and the third is ω times the second.

Example 1 A particle oscillates in a horizontal line about a fixed point O such that its distance, x, from O at time t is given by $\mathbf{x = 5\sin 2t}$.

Find (a) the velocity and acceleration when $x = 3$,

(b) x and $\dot{x}$ when $\ddot{x} = 1$.

(a) From the given equation we know that $a = 5$ and $\omega = 2$. And we are given $x = 3$, so the chart is

$$\begin{array}{r|lll} & - & 3 & 5 \\ 2t & \dot{x} & - & - \\ & - & -\ddot{x} & - \end{array}$$

Completing the triple in the first line we can put ± 4 down, and the last elements of the other two triples are found by multiplying 5 by ω and then by ω again, giving 10 and 20:

$$\begin{array}{r|lll} & \pm 4 & 3 & 5 \\ 2t & \dot{x} & - & 10 \\ & - & -\ddot{x} & 20 \end{array}$$

It now becomes clear that since the second triple is double the first (as 10 is double 5), $\dot{x} = \pm 8$.

We now have the first two triples, and $-\ddot{x}$ in the last line:

$$\begin{array}{r|lll} & \pm 4 & 3 & 5 \\ 2t & \pm 8 & 6 & 10 \\ & - & -\ddot{x} & 20 \end{array}$$

Since the bottom triple is double the middle one this gives us $-\ddot{x} = 12$ and so $\ddot{x} = -12$. We could also obtain $-\ddot{x}$ from the first triple by doubling 3 twice.

(b) We have $\ddot{x} = 1$ so this can be put in the middle of the last triple, and this triple can then be completed:

	-	x	5
2t	x	-	10
	$\pm\sqrt{399}$	-1	20

Now we can see that we have to divide $\pm\sqrt{399}$ by 2 to get x, and divide -1 by 4 to get x. So $x = -\frac{1}{4}$ and $x = \pm\frac{\sqrt{399}}{2}$

Note here that since we are dealing with triples the last element of any triple gives a maximum and minimum value.
For example, from the above chart: $x_{max} = 5$,

$x_{max} = 10$,

$\ddot{x}_{min} = -20$.

Example 2 A particle is describing linear S.H.M. with a period of 4 seconds. Given that the greatest speed is 3m/s find the amplitude of the path and the speed of the particle when it is $\frac{2}{\pi}$ m from the centre.

If the period is 4 seconds then $\omega = \frac{2\pi}{4} = \frac{\pi}{2}$.
So if a is the required amplitude we have:

-	-	a
-	-	$\frac{a\pi}{2}$
-	-	-

But since $\frac{a\pi}{2}$ is the maximum speed, we have $\frac{a\pi}{2} = 3$ and $a = \frac{6}{\pi}$.

Inserting this value and the given $\frac{2}{\pi}$ we now have:

-	$\frac{2}{\pi}$	$\frac{6}{\pi}$
-	-	3
-	-	-

Seeing the ratio 1:3 in the first line we can now complete the second triple:

-	$\frac{2}{\pi}$	$\frac{6}{\pi}$
$\pm\sqrt{8}$	1	3
-	-	-

So $x = \pm\sqrt{8}$.

Example 3 A particle oscillating with linear S.H.M. has an amplitude of 3m. When $\sqrt{5}$ m from the centre of oscillation its speed is $10ms^{-1}$. Find the maximum speed of the particle.

The chart when $x = \sqrt{5}$ m is:

-	$\sqrt{5}$	3
10	-	-
-	-	-

and the first element of the first triple is therefore ± 2.

Since this gives a ratio between the first two triples of 1:5 we can set up a chart for $x = 0$ (when the speed of the particle is maximum):

	3	0	3
	15	0	15
	-	-	-

The maximum speed is therefore clearly $15ms^{-1}$.

<u>Example 4</u> Two particles P and Q perform linear S.H.M. with the same parameters as Example 1, as shown in the chart below. When P first reaches B, where OB = 3m, Q passes through O, the mean position.

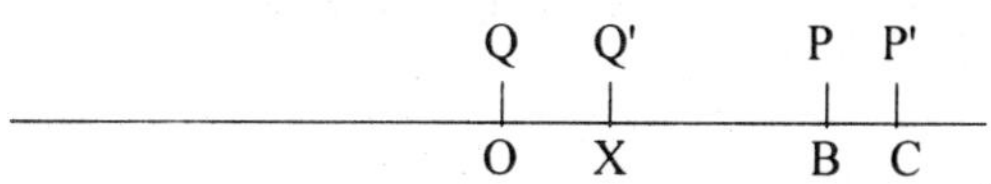

	-	x	5
2t	$\dot{x}$	-	10
	-	$-\ddot{x}$	20

Find (a) the distance travelled by Q when P first reaches C, where OC = 4m,

(b) the time for P to cover the distance BC.

We know that $\omega = 2$ and $a = 5$.

(a) The two "distance" triples for P at B and at C are (4),3,5 and (3),4,5 respectively.

Subtracting these two triples:

3	4	5	
4	3	5	−
24	7	25	

gives a triple whose angle gives the time taken for P to travel from B to C, and Q to travel from O to X, say, where OX is the required distance. We can now divide this triple by 5, so that its last element is equal to a. This gives the equal triple 4.8, 1.4, 5 which can be thought of as a "distance" triple in which $x = 1.4$. So <u>OX = 1.4m</u>.

(b) We need the angle in the triple 24,7,25. This is a known triple so we can put $2t = 0.2838$, and so <u>$t = 0.1419s$.</u>

For problems where Simple Harmonic Motion starts at an extremity of the motion it is sometimes useful to use a system based on $x = a\cos \omega t$ rather than $x = a\sin \omega t$.

The general chart in this case would be:

$$\begin{array}{r|ccc} & x & - & a \\ \omega t & - & -\dot{x} & a\omega \\ & -\ddot{x} & - & a\omega^2 \end{array}$$

Distances are measured from the mean position in both systems, but in this second system time is measured from an extremity.

<u>Example 5</u> A particle describes linear S.H.M. with period 12 seconds between two points A and B. Given that the particle is 0.5m from A after 2 seconds find its velocity 4 seconds after leaving A.

A period of 12 seconds means that $\omega = \frac{\pi}{6}$.
It would be convenient in this example to use the second system with time being measured from position A.

Then the chart for t = 2 would be:

$$\begin{array}{r|ccc} & 0.5 & - & a \\ \frac{\pi}{3} & - & - & - \\ & - & - & - \end{array}$$

Since the triple for $\frac{\pi}{3}$ is $1, \sqrt{3}, 2$ we can see by proportion that $a = 1$.

Now consider the chart for $t = 4$:

$$\begin{array}{r|ccc} & - & - & - \\ \frac{2\pi}{3} & - & -\dot{x} & \frac{\pi}{6} \\ & - & - & - \end{array}$$

The triple for $\frac{2\pi}{3}$ is $-1, \sqrt{3}, 2$ and again by proportion we find that $\dot{x} = -\frac{\pi\sqrt{3}}{12}$.

As mentioned earlier the standard formulae for Simple Harmonic Motion can be established from the initial general chart:

$$\begin{array}{r|ccc} & - & x & a \\ \omega t & \dot{x} & - & a\omega \\ & - & -\ddot{x} & a\omega^2 \end{array} \qquad \begin{array}{c} (1) \\ (2) \\ (3) \end{array}$$

Multiplying (1) by ω^2 so that the elements may be equated to those in (3) we get $\ddot{x} = -\omega^2 x$.
And multiplying (1) by ω and equating the first elements in (1) and (2) we get $\dot{x} = \omega\sqrt{a^2 - x^2}$.
Other formulae, not normally used, are also available by equating other elements.

Exercise A

1. A particle oscillates in a horizontal line about a fixed point such that its distance, xm, from the fixed point at time t seconds is given by $x = 2\sin 3t$.
Find the velocity and acceleration when x=1m.

2. A particle oscillates in a horizontal line about a fixed point such that its distance, xm, from the fixed point at time t seconds is given by $x = 2\sin 3t$.

Find its distance from the fixed point and its acceleration when $x = 3$m/s.

3. If the equation of motion is $x = 5\sin 2t$, with the same units as above find when the particle is 4m from the origin and when $x = 8$.

PROJECTILES

A particle moving under a constant gravitational force would have an "acceleration triple":

A) 0, $-g$, -

meaning that the acceleration, a, is zero in a horizontal direction and equal to $-g$ (g is the acceleration due to gravity) in a vertically upwards direction.

Integrating this with respect to time gives the "velocity triple":

V) w , $u-gt$, -

where w is the initial velocity in the horizontal direction and u is the initial velocity in the vertical direction.

Integrating again we obtain the "distance triple":

S) wt , $ut-\frac{1}{2}gt^2$, -

assuming that time is measured from when the particle was at the origin of coordinates.

These three triples contain the usual formulae for this type of motion in a compact and accessible form that is easily derived from first principles.

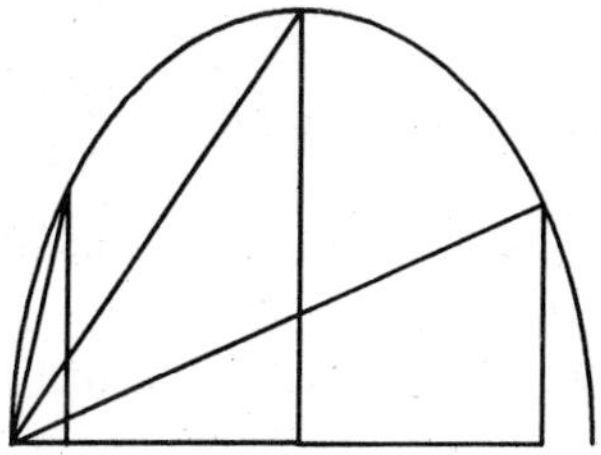

distance triples

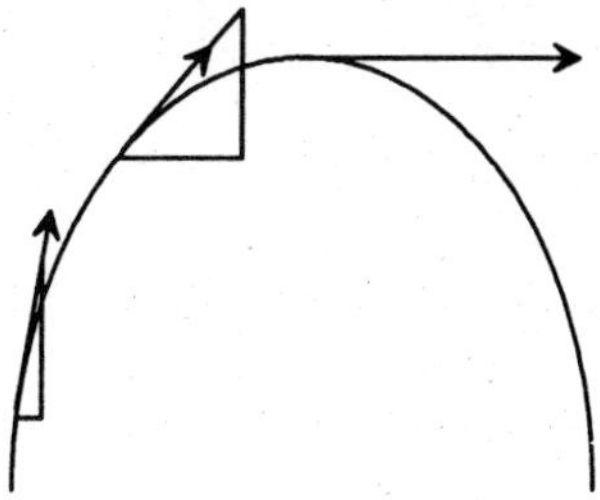

velocity triples

Putting the three triples together we may write:

A	0	$-g$	g
V	w	$u-gt$	speed at time t
S	wt	$ut-\frac{1}{2}gt^2$	distance from origin at time t

in which we complete the three triples, and in which, we use the letters A, V and S to represent the angles in these triples.

Example 6 A particle is projected at $200ms^{-1}$ at an angle of elevation of $\tan^{-1}(4/3)$.
Find (a) its greatest height,
(b) its range on a horizontal plane,
(c) its speed after 7 seconds,
(d) when its velocity is at $\tan^{-1}(5/12)$ to the horizontal
(e) after how long the velocity is perpendicular to its original velocity.

Using $g = 10ms^{-2}$ and finding the components of the initial velocity to be $\frac{3}{5}\times 200 = 120$ and $\frac{4}{5}\times 200 = 160$ the general chart here is:

A	0	-10	-
V	120	$160-10t$	-
S	120t	$160t-5t^2$	-

a) The greatest height will be when the vertical component of the velocity is zero.
That is $160 - 10t = 0$. So $t = 16s$.
Putting this value into the vertical component of the distance triple we get 1280m as the greatest height.

b) The range will be twice the base of the distance triple just discussed, so it is 3840m.

c) The speed is given by the third element of the velocity triple. Putting $t = 7$ we get 120 and 90 as the first two elements of this triple so that the third element is $150ms^{-1}$.

d) Here the velocity triple given is 12,5,13 and this is equated to the general velocity triple.
So we can write $\frac{120}{160-10t} = \frac{12}{5}$, from which we find that $t = 11s$.

e) A triple perpendicular to 120, 160, - is 160, –120, -, but since the first element is always 120 we multiply this through by $\frac{3}{4}$ to obtain 120, –90, - as the perpendicular velocity triple.
Comparing this with the general velocity triple we can write
$160 - 10t = -90$.
From which $t = 25s$.

Example 7 A particle is projected with a velocity of $40\sqrt{2}$ m/s at an angle of 45° from the edge of a cliff. The particle is sighted from the point of projection when at its greatest height. Find the time that elapses from then until its angle of elevation, as seen from the point of projection, has decreased by the angle in the triple 4,3,5.

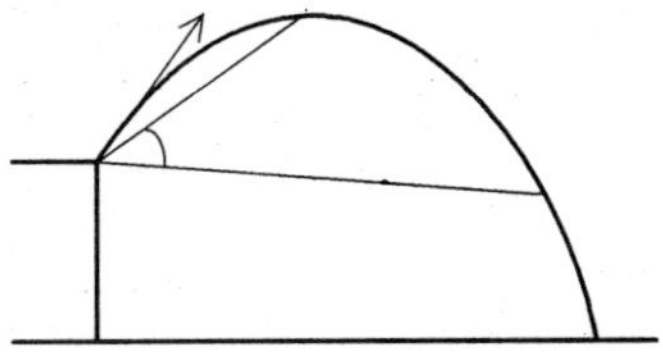

Since the triple for 45° is $1, 1, \sqrt{2}$, the initial velocity triple is $40, 40, 40\sqrt{2}$.
So the general chart is:

A	0	−10	-
V	40	40 − 10t	-
S	40t	$40t - 5t^2$	-

At the greatest height we have $40 - 10t = 0$, so $t = 4$s.
The distance triple at this time is therefore 160, 80, - , which cancels to 2, 1, -.
And from this we should subtract the 4,3,5 triple:

2	1	-	
4	3	-	−
11	−2	-	

We see from this answer that the particle is seen below the horizontal.
At this time its direction is given by 11, −2, -, though these are not the right distances.
This is not a problem however as we are going to equate the ratio of distances to that in the general distance triple: $\frac{40t}{40t-5t^2} = \frac{11}{-2}$ from which $t = \frac{104}{11}$ s.

Example 8 A particle is projected at 30m/s at an angle of $\arctan\frac{3}{4}$ to the horizontal. When has the direction of its velocity changed by the angle in the triple 12,5,13?

The general chart is:

A	0	−10	-
V	24	18 − 10t	-
S	24t	$18t - 5t^2$	-

Here the original direction is given by 4,3,5 and this is to be reduced by 12,5,13.
We therefore subtract these triples:

4	3	5	
12	5	13	−
63	16	65	

Since this is the final velocity direction we need to find the value of t for which it is equal to the general velocity triple: $\frac{63}{16} = \frac{24}{18-10t}$ from which $t = \frac{25}{21}$ s.

There are various advantages in using the triple system here, including:

(a) the use of distance triples and velocity triples, and the significance of their last element,
(b) the use of addition and subtraction of triples,
(c) the structuring of the various equations in a single chart, which becomes even more useful when two or more projectiles occur in the same problem (see exercise below).

Exercise B (use $g = 10m/s^2$)

1. A particle is projected at $\arctan\frac{3}{4}$ with a speed of 100m/s.
   Find (a) its speed and position after 2 seconds,
   (b) the change in direction of the projectile's motion from t = 2 to t = 3s,
   (c) the change in the angle of elevation of the projectile from t = 2 to t = 3s as seen from its point of projection.

2. Two particles are projected in the same vertical plane, at the same instant, from the same point and with a speed of 20m/s. They are projected on horizontal ground at angles of 30° and 60°.
   (a) Write down the general chart for each projectile.
   (b) Find the position of each projectile when the lower one hits the ground.

3. Two particles A and B are projected in the same vertical plane, at the same instant and from the same point, one with an initial speed of 26m/s at an angle of $\arctan\frac{5}{12}$ and the other with an initial speed of 30m/s and at an angle of $\arctan\frac{3}{4}$. Find the position of B when A is at its greatest height and their distance apart when t = 2s.

FORCES IN EQUILIBRIUM

Various methods are available for solving problems with forces in equilibrium including resolution of forces in perpendicular directions, the triangle of forces and Lami's theorem.

If two of three forces acting on a body are perpendicular then a triple of forces results and simple proportion can provide the solution. And, as Example 12 shows, even when none of the forces are perpendicular (or when more than three forces are involved) this triple method can still be used.

Example 9 An object of weight 20N is attached to one end of a string AC 2m long. The other end, A, of the string is fixed and a horizontal force is applied at C so that the system hangs in equilibrium as shown, with C 120cm from the vertical through A. Find the magnitude of the force and the tension in the string.

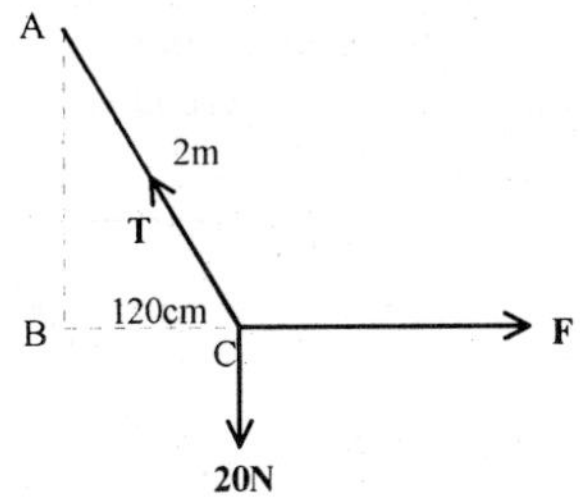

The triangle ABC shown is the triple 120 - 200 which is equal to 3,4,5. Since the three forces are parallel to the sides of this triangle ABC also serves as a triple of forces.

So the triple of forces **F** **20** **T**
is equal to **3** **4** **5.**

And since the middle element of the first is 5 times the middle element of the second we can see that F = 15N and T = 25N.

Example 10 Wires of length 8m and 6m are attached at each end of a beam 10m long. The other ends are attached to a third wire from which a weight of 20N hangs. Find the tensions in the three wires.

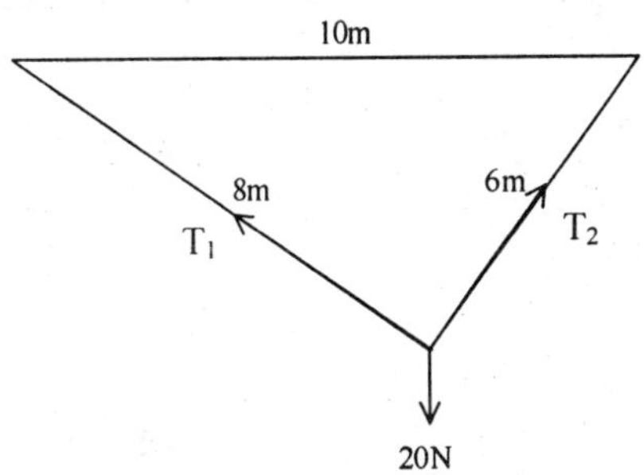

The diagram shows the arrangement and it is clear that 6, 8, 10 is a triple so that the tensions T_1 and T_2 are perpendicular. $T_3 = 20N$, so the hypotenuse of the triple of forces is 20N.

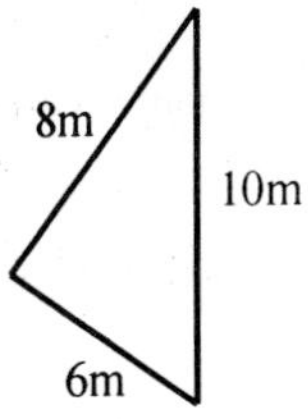

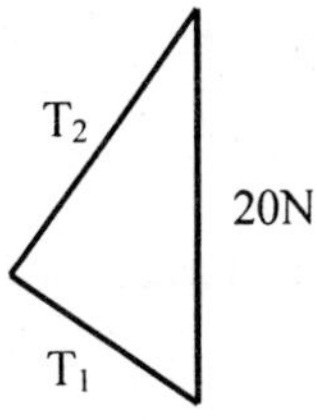

6	8	10
T_1	T_2	20

It follows that the three forces are 12N, 16N and 20N respectively.

Example 11 A rod AE of weight 21N and length 10m is hinged at A and is supported as shown by a horizontal wire DE where D is 7m vertically above A. Find the tension in the wire.

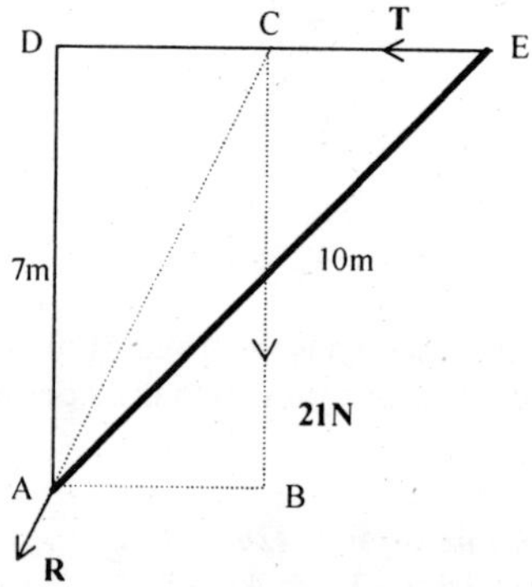

The three forces (tension T, weight 21N, and the hinge reaction R) acting on the rod are in equilibrium and so must all pass through a point. This point is C, the mid point of DE.

Thus triangle ABC can represent a triple of forces.

From triangle ADE we see that DE = $\sqrt{51}$ and that AB is therefore $\frac{\sqrt{51}}{2}$.

So the triple of forces is	T	21	R
and its dimensions are	$\frac{\sqrt{51}}{2}$	7	-

The ratio of the middle elements here clearly gives $T = \frac{3\sqrt{51}}{2}$ N.

Example 12 A particle of weight 50N rests on a smooth slope inclined at 30° to the horizontal. It is supported in this position by a force F which is inclined at 60° to the horizontal. Find F.

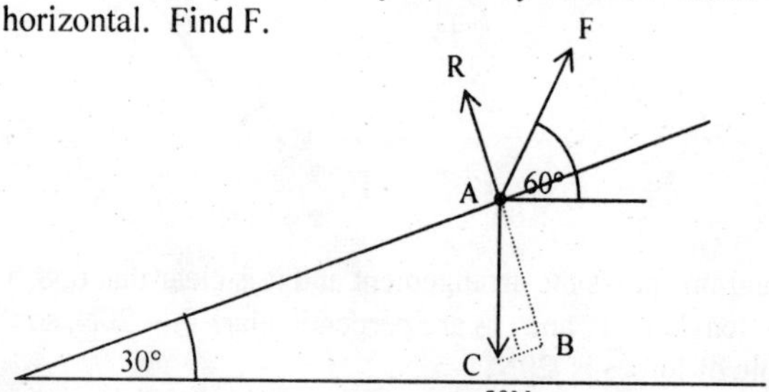

The best way to solve this is probably to resolve forces parallel to the plane, but the triple method can be used by resolving F into components along and perpendicular to the plane.

The triangle ABC can be viewed as a triple of forces in which

the triple of forces is	$R + \frac{F}{2}$	$\frac{F\sqrt{3}}{2}$	50
and the dimensions are	$\sqrt{3}$	1	2

Looking at the ratios in the last two columns $\frac{F\sqrt{3}}{2} = 25$ and $F = \frac{50}{\sqrt{3}}$ N.

Exercise C

1. A weight of 50N is suspended by a string attached to a fixed point. The weight is pulled aside by a force F until the string is inclined at 30° to the vertical. Find F and the tension in the string.

2. A sphere of weight 20N and radius r rests against a smooth vertical wall supported by a string of length r attached to a point on the sphere and to the wall. Find the tension in the string.

3. A ladder of weight 30N, on rough horizontal ground, leans against a smooth vertical wall at an angle of $\arctan(\frac{12}{7})$ to the horizontal. Find the magnitude of the reaction at each end of the ladder.

4. A uniform rod AB of length 13m and weight 6N is hinged at A to a vertical wall. A horizontal force F is applied at B so that B is 5m vertically below A. Find F and the magnitude of the force at the hinge.

WORK DONE BY A FORCE AND MOMENT OF A FORCE

Example 13 Given $\mathbf{F} = 5\mathbf{i} + 4\mathbf{j}$ and $\mathbf{r} = 3\mathbf{i} + 2\mathbf{j}$
find a) the work done when the force **F** moves through the vector **r**,
and b) the moment about the origin of the force **F** acting through the point with position vector **r**.

$$\begin{array}{rrr} \mathbf{F} = & 5\mathbf{i} & + \ 4\mathbf{j} \\ \underline{\mathbf{r} =} & \underline{3\mathbf{i}} & \underline{+ \ 2\mathbf{j}} \\ & \underline{23} & \underline{2} \end{array}$$

Using the method of triple subtraction we get

$5 \times 3 + 4 \times 2 = 23\text{J}$ as the work done,

and $4 \times 3 - 5 \times 2 = 2\text{Nm}$ for the moment.

When quadruple subtraction is defined later this method can be extended to the 3-dimensional case (see Chapter 15).

13 THE TRIPLE METHOD

Before extending triples into the 3rd dimension in the next chapter it will be useful to examine the triple method and discuss its advantages.
The term "triple method" means expressing a problem in triple form and manipulating the triple(s) to solve the problem.

- The right-angled triangle is a fundamental unit in geometry and of course has three sides and three angles. The triple method works with the sides of triangles and in manipulating and combining triples using their sides we are also manipulating and combining their angles. This means we can avoid mixing incommensurable units- sides and angles.

- A triple can represent many things: e.g. a triangle (of distances, forces, velocities, etc.), an angle, a gradient, the position of a point.

- Where desired angular answers can also be given in terms of the sides of a triangle. For example A) –2, 1, - describes an angle exactly. And an impression of the angle is more easily obtained from this than from $A=\cot^{-1}(-2)$. In general triples give a clearer understanding of what is being dealt with.

- The calculation of sines, cosines, tangents etc. and square roots, that occur in the solution of some problems, can be avoided by the triple method.

- The triple method unites in a single method a variety branches of mathematics. There is no similarity between the various formulae of trigonometry, coordinate geometry, transformations etc but the simple single triple method can be applied in them all.

- The triple notation is compact and comprehensive. The angle and the three sides are included and the value of any two of these defines a particular triple. If a value is not known or not required a letter or a dash can be inserted. Algebraic operations can also be carried out on general triples.

- The triple notation also combines the usual Cartesian and polar notations. This means the notation is just as effective with polar equations as with Cartesian (see below).

❖ The simple Vertical and Crosswise pattern makes the addition and subtraction of triples easy.

❖ The word abbreviations sin, cos, tan, cosec, sec and cot, which confuse many students, can be avoided by the triple method.

RANGE OF APPLICATION

There are various types of triple: perfect, non-perfect, prime, constant and variable triples. We have complementary and supplementary triples, equal triples, vanishing triples, 2 and 3 dimensional (and higher order) triples.

Triples can be added, subtracted, doubled, halved, differentiated and integrated.

Triple elements and their code numbers can also be imaginary or complex.

Triples can be used in a large range of areas of mathematics as demonstrated in this book and there are certainly many more applications to be found. For example, the relations between triples, vectors and complex numbers could be studied further, including eigenvectors and De moivre's theorem. There may also be applications in wave theory, AC circuit theory, conformal mappings, quaternions, astronomy etc. There are certainly applications in computing: the use of integers instead of trigonometrical functions and square roots can considerably increase the speed of computations.

Below we give a few examples of different types of triples.

Triple form	**Conventional form**	
A)1, -, 3	$\cos A = \frac{1}{3}$	trig. equation
x)3x, -, 1	$\cos x = 3x$	a transcendental equation
x–a, y–b, 3	$(x-a)^2 + (y-b)^2 = 9$	equation of a circle
t)2x, 3y, 12	$\frac{x^2}{36} + \frac{y^2}{16} = 1$ or x=6cost, y=4sint	equation of an ellipse
4, 3y, 2x	$\frac{x^2}{4} - \frac{9y^2}{16} = 1$	equation of a hyperbola
2t) x, -, 10	$x = 10\cos 2t$	velocity equation
A) r, -, 2a	$r = 2a\cos A$	polar equation of a circle
A) r–a, -, a	$r = a(1 + \cos A)$	polar equation of a cardioid

Furthermore the applications shown in this book are based on the triple addition defined in Chapter 2. Alternative ways of combining or manipulating triples could lead to other useful applications.

DERIVING THE CONVENTIONAL FORMULAE

In every case it is easy to convert the triple form above to the conventional form or vice versa. Similarly it is also easy to obtain conventional formulae using the triple method.

For example to rotate a point (x,y) through an angle A about the origin the matrix

$\begin{pmatrix} \cos A & -\sin A \\ \sin A & \cos A \end{pmatrix}$ is generally used and this multiplies $\begin{pmatrix} x \\ y \end{pmatrix}$

$$\begin{pmatrix} \cos A & -\sin A \\ \sin A & \cos A \end{pmatrix}\begin{pmatrix} x \\ y \end{pmatrix} = \begin{pmatrix} x\cos A - y\sin A \\ x\sin A + y\cos A \end{pmatrix} \qquad (1)$$

The student is expected (as with all the other formulae) to know the formula or to look it up. And for rotations the student also needs to understand matrix multiplication.

The triple method can give this formula too:

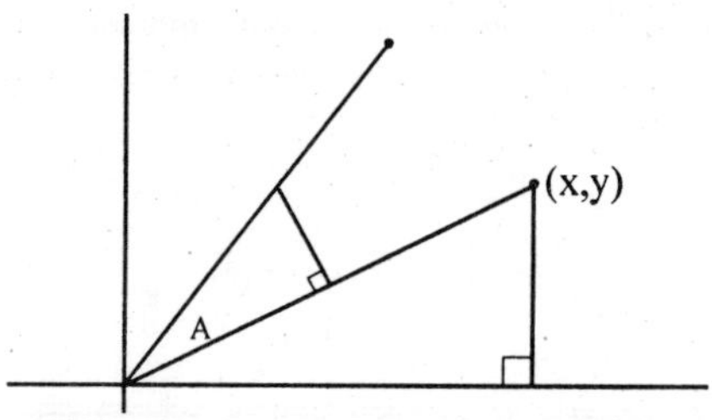

The point (x,y) is being rotated by an angle A about the origin.
We therefore add the triples just as in Chapter 4:

$$\begin{array}{ccc} x & y & \sqrt{x^2+y^2} \\ \cos A & \sin A & 1 \\ \hline x\cos A - y\sin A & x\sin A + y\cos A & \sqrt{x^2+y^2} \\ \hline \end{array} \quad +$$

We see that the first two elements of the answer are the components of the right-hand side of equation (1) above.

Formulae which the triple method can replace (as shown in this book) include:

From trigonometry $\cos^2 A + \sin^2 A = 1$

$1 + \tan^2 A = \sec^2 A$

$1 + \cot^2 A = \operatorname{cosec}^2 A$

$\sin 2A = 2\sin A\cos A$

$\cos 2A = \cos^2 A - \sin^2 A = 2\cos^2 A - 1 = 1 - 2\sin^2 A$

$\tan 2A = \frac{2\tan A}{1-\tan^2 A}$

$\sin(A+B) = \sin A\cos B + \cos A\sin B$

$\sin(A-B) = \sin A\cos B - \cos A\sin B$

$\cos(A+B) = \cos A\cos B - \sin A\sin B$

$\cos(A-B) = \cos A\cos B + \sin A\sin B$

$\tan(A+B) = \frac{\tan A+\tan B}{1-\tan A\tan B}$

$\tan(A-B) = \frac{\tan A-\tan B}{1+\tan A\tan B}$

If $t=\tan\frac{1}{2}A$, $\sin A=\frac{2t}{1+t^2}$, $\cos A = \frac{1-t^2}{1+t^2}$, $\tan A = \frac{2t}{1-t^2}$

$2\sin A\cos B = \sin(A+B) + \sin(A-B)$

$2\cos A\cos B = \cos(A+B) + \cos(A-B)$

$2\sin A\sin B = \cos(A-B) - \cos(A+B)$

$\sin A + \sin B = 2\sin\frac{A+B}{2}\cos\frac{A-B}{2}$

$\sin A - \sin B = 2\cos\frac{A+B}{2}\sin\frac{A-B}{2}$

$\cos A + \cos B = 2\cos\frac{A+B}{2}\cos\frac{A-B}{2}$

$\cos A - \cos B = -2\sin\frac{A+B}{2}\sin\frac{A-B}{2}$

Plus the corresponding formulae for the hyperbolic functions.

From coordinate geometry $d = \frac{|ah+bk+c|}{\sqrt{a^2+b^2}}$ (distance of the point (h,k) from the line ax+by+c=0)

$\tan A = \frac{m_1-m_2}{1+m_1m_2}$ (angle between two straight lines)

From transformational geometry

$\begin{pmatrix}\cos A & -\sin A\\ \sin A & \cos A\end{pmatrix}$ (the 2-dimensional rotation matrix)

$\begin{pmatrix}\cos 2A & \sin 2A\\ \sin 2A & -\cos 2A\end{pmatrix}$ (for reflecting in the line whose equation is y=xtanA)

Simple Harmonic Motion $\ddot{x} = -\omega^2 x$

$\dot{x}^2 = \omega^2\left(a^2 - x^2\right)$

$x = a\sin\left(\omega t + \varepsilon\right)$

Projectile motion $v = u + ft$

$s = ut + \frac{1}{2}ft^2$

The triple method offers a simple alternative to all these formulae. The extent of the applications of triples shown in this book is still being explored and further applications are likely to arise.

It is remarkable that the triples can be applied in so many different areas of mathematics.

TWO COMPARISONS OF THE CONVENTIONAL AND TRIPLE METHODS

Finally we can give two illustrations of the effectiveness of the triple method.

1) To find the general solution of the equation $2\sin x - \cos x = 2$.
The student is usually encouraged to use the standard formula:
$R\sin(x - a) = R\sin x\cos a \; R\cos x\sin a$ and by equating coefficients find the values of R and a for which $2\sin x - \cos x = R\sin(x-a)$.

This involves using a calculator to find a and making an approximation for the value of a. This leads to the equation $\sqrt{5}\,\sin(x - 0.4636) = 2$, if a is measured in radians.

Then we must divide 2 by $\sqrt{5}$, find the inverse sine of the result, express the answer in general terms and add the 0.4636 to the right-hand side.

The calculator is used at least three times and the final answers are approximate.

This example was used in a book on 'A' level mathematics and follows the method shown above.
The triple method gives the calculation in a simple compact form and gives the exact answers, describing the angles as triples:

$$\begin{array}{rrrrl} A: & x & y & 1 & \\ B: & -1 & 2 & \sqrt{5} & - \\ \hline A-B: & 2 & 1 & \sqrt{5} & + \\ \hline A: & -4 & 3 & 5 & \\ A: & 0 & 5 & 5 & \\ \hline \end{array}$$

2) To find $\sqrt{15+8i}$.

The usual method here is to call the required result $a + bi$, square both sides of the equation thus formed and equate real and imaginary parts to get $a^2 - b^2 = 15$ and $2ab = 8$.

Substituting from the second equation into the first leads to a quartic equation which is then factorised to give two quadratic equations. This gives the two values of a and the corresponding values of b can then be found, giving $\sqrt{15+8i} = \pm(4+i)$

This example was also done in an 'A' level textbook and takes 17 lines of working.

By contrast the triple method allows the answer to be written straight down: the triple 15, 8, 17 is mentally completed and its code numbers give the answer: $\sqrt{15+8i} = \pm(4+i)$

These are just two examples of many that could be chosen to illustrate the simplicity, directness and speed of the triple method.

14 QUADRUPLES

INTRODUCTION

The triples idea developed so far in this book can be further extended in various ways. We could, for example, study integral-sided triangles containing not a right angle but an angle of 60° or 120° instead. The triangle with sides 3, 5, 7 contains an angle of exactly 120°:

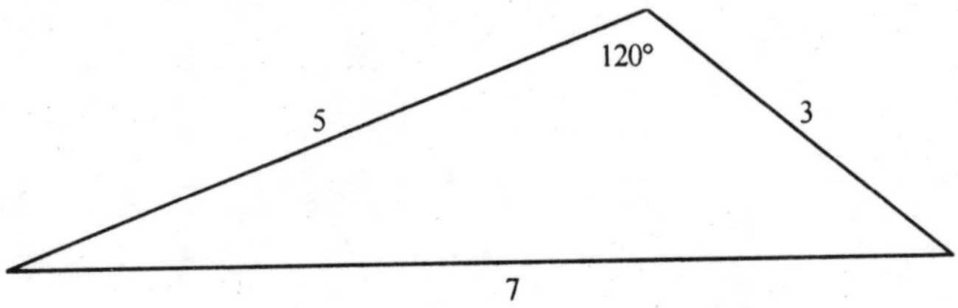

These triangles can be combined in various ways and are similar in many respects to the perfect triples.

Alternatively, we may consider triangles with integral sides which are almost right angles. The "near triple" 7, 11, 13 mentioned in Chapter 1 is not right-angled as $7^2 + 11^2 = 170$ and $13^2 = 169$ but it is very nearly right-angled.
The angle deficiency (see Chapter 7) $7^2 + 11^2 - 13^2 = 1$ so all the angles are acute.
These triangles also have many fascinating properties and are closely linked to the perfect triples.

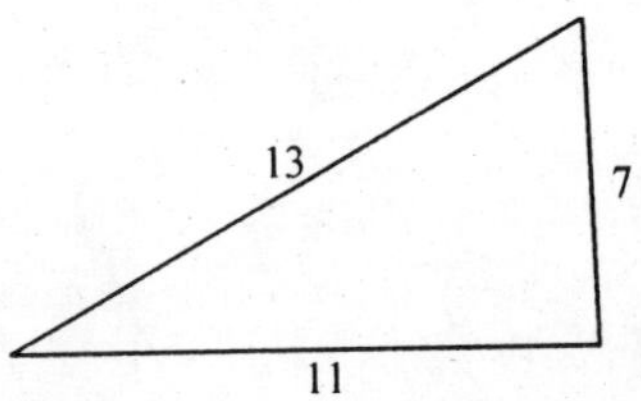

A third way of extending triples is to develop a 3-dimensional triple that can define a direction in 3-dimensional space, just as ordinary triples can define a direction in 2-dimensional space. This leads to the notion of "quadruples".

These will be defined and developed here along similar lines to the triples, and we will see that they have very useful applications in various types of 3-dimensional problems.

6, 3, 2, 7 is a **perfect quadruple** as $6^2 + 3^2 + 2^2 = 7^2$ ($36 + 9 + 4 = 49$).
Any four numbers that have this property is a quadruple, and if the numbers are also rational as above it is a perfect quadruple.
So if x, y, z, r is a quadruple then $x^2 + y^2 + z^2 = r^2$, and conversely.

The quadruple 6, 3, 2, 7 describes the point P(6,3,2) in 3-dimensional space which is exactly 7 units from the origin O:

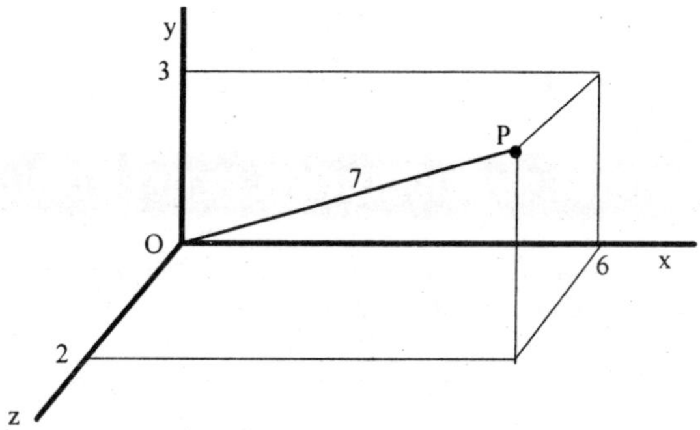

Other examples of perfect quadruples are:

1	70	70	99
2	–1	2	3
0	3	4	5
49	64	8	81

Note that the third example here: 0,3,4,5 is just a triple in the y-z plane.
If any element of a quadruple is zero the quadruple is also a triple, so all triples are contained within the 3-dimensional quadruples.

Quadruples can be represented usefully by a pyramid composed of four right-angled triangles:

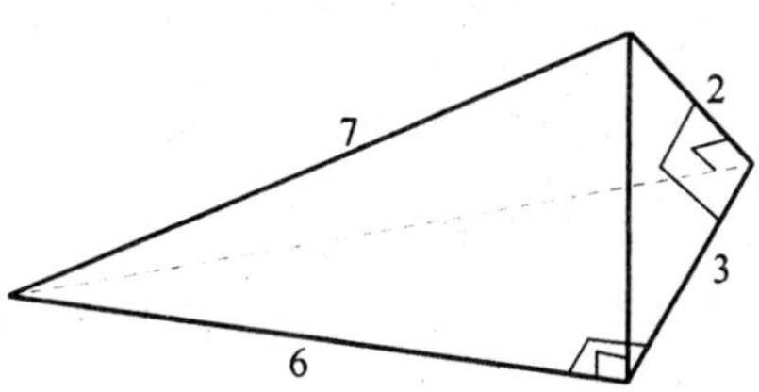

We will discuss later what we mean by the angle in a quadruple.

QUADRUPLE GENERATORS

The formula $c^2 - d^2 + e^2$, $2cd$, $2de$, $c^2 + d^2 + e^2$ generates perfect quadruples for all rational values of c,d,e.
c, d, e are the **code numbers** of the quadruple.
For example, for code numbers **3,1,2** the above formula gives **12,6,4,14 = 6,3,2,7**.

There are a number of other formulae that also generate perfect quadruples. And in addition perfect quadruples can be produced:

a) from triples (compare the generating formulae),
b) by the addition of two triples (see Pages 147 and 151),
c) by the addition of a triple and a quadruple (see Page 149),
d) by the addition or subtraction of two quadruples (see Pages 152,3).

OBTAINING THE CODE NUMBERS OF A PERFECT QUADRUPLE

The code numbers of the quadruple **6, 3, 2, 7** are **3, 7–6, 2 = 3, 1, 2**.
That is, in general, the code numbers of **x, y, z, r** are **y, r–x, z**.

You may like to check that the code numbers of the four quadruples given on Page 133 are:

5, 7, 5, (any common factor can be divided out)
–1, 1, 2,
3, 5, 4
and **8, 4, 1**.

So the code numbers **1, 1, 3** generate the quadruple **9, 2, 6, 11**,
and the code numbers of **9, 2, 6, 11** are **2, (11–9), 6 = 1, 1, 3**.

Similarly code numbers **c,d,e** generate $c^2 - d^2 + e^2$, $2cd$, $2de$, $c^2 + d^2 + e^2$.
And the code numbers of $c^2 - d^2 + e^2$, $2cd$, $2de$, $c^2 + d^2 + e^2$ are $2cd$, $2d^2$, $2de$ = **c, d, e**.

THE COORDINATE AXES

The quadruple **1,0,0,1** describes the positive x-axis.
The six fundamental directions are given below with their code numbers:

	quadruple	code numbers
+ve x-axis	**1 0 0 1**	**1 0 1***
–ve x-axis	**–1 0 0 1**	**0 1 0**
+ve y-axis	**0 1 0 1**	**1 1 0**
–ve y-axis	**0 –1 0 1**	**1 –1 0**
+ve z-axis	**0 0 1 1**	**0 1 1**
–ve z-axis	**0 0 –1 1**	**0 –1 1**

* For this quadruple the code numbers cannot be found by the method described in the last section. It is the only exception however.

QUADRUPLE SUBTRACTION

We define quadruple subtraction as follows:

$$\begin{array}{ccccc} x & y & z & & r \\ \underline{X} & \underline{Y} & \underline{Z} & & \underline{R} \quad - \\ \underline{xX+yY+zZ,} & \underline{yX-xY,} & \underline{zX-xZ,} & \underline{zY-yZ,} & \underline{rR} \end{array} \qquad (1)$$

For example:

$$\begin{array}{ccccc} 2 & 3 & 6 & & 7 \\ \underline{2} & \underline{2} & \underline{1} & & \underline{3} \quad - \\ \underline{16} & \underline{2} & \underline{10} & \underline{9} & \underline{21} \end{array}$$

The result is a "quintuple": $16^2 + 2^2 + 10^2 + 9^2 = 21^2$.
That the difference of any two quadruples is a quintuple (that is, that the sum of the squares of the first four elements is equal to the square of the 5th) is easily proved from (1) above.

We define quadruple subtraction in this way because
(a) if one element, say the 4th, in (1) is zero (i.e. if zY=yZ) we obtain a perfect quadruple; as will be seen later this leads to very interesting and useful applications,
(b) if the middle three elements in (1) are combined into one element the result is a triple with useful properties (see the next chapter).

Note that the 1st element and the 2nd, 3rd and 4th elements in (1) are related to the scalar and vector products respectively of vector algebra. There is also a connection here with multiplication of quaternions.

COMPARATIVE DENSITIES OF PERFECT TRIPLES AND PERFECT QUADRUPLES

There are 34 perfect triples which have all their elements less than 100: 17 of the form a,b,c and 17 of the form b,a,c. So 34 distinct directions are defined in the first quadrant. There are 348 quadruples of the form a,b,c,d and 17 of the form a,a,b,c (in which all the elements are below 100). But the quadruples describe directions in 3-dimensional space and the triples in 2-dimensional space, so are the quadruples more numerous than the triples?
Consider a quadrant of a circle of unit radius: we have 34 triples spread over an arc of length $\frac{\pi}{2}$, so a similar density would be 34^2 triples over an area of $\frac{\pi^2}{4}$. But the surface area of an octant of a sphere of unit radius is $\frac{\pi}{2}$, so over this area we would expect $\frac{34^2}{\frac{\pi}{2}} \approx 736$ distinct directions. Now the 348 quadruples of the form a,b,c,d gives 348×6 by rearranging the elements, and the 17 of the form a,a,b,c give 17×3. The total is 2139 quadruples in the first octant, making them about 3 times more numerous than the triples.

15 APPLICATIONS OF QUADRUPLES

COORDINATE GEOMETRY

Just as triples can simplify work in 2-dimensions quadruples and triples have many useful applications in 3-dimensional space.

The usual Cartesian and vector equations of a line and a plane are:

Line: $\frac{x-p}{a} = \frac{y-q}{b} = \frac{z-r}{c}$ and $\mathbf{r} = p\mathbf{i} + q\mathbf{j} + r\mathbf{k} + l(a\mathbf{i} + b\mathbf{j} + c\mathbf{k})$

where (p,q,r) are the coordinates of a point on the line and P is the point P(a,b,c) such that OP is in the direction of the line.

Plane: $ax + by + cz = d$ and $\mathbf{r} \sqcup (a\mathbf{i} + b\mathbf{j} + c\mathbf{k}) = d$

where P is the point P(a,b,c) such that OP is perpendicular to the plane.

<u>Example 1</u> Find a triple for angle AOB where A is A(8,4,1) and B is B(6,3,2).

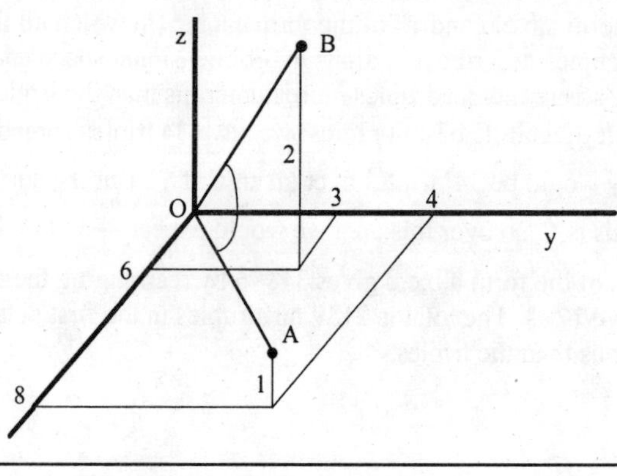

We complete the two quadruples (i.e. insert the fourth element) and subtract them using formula (1) from Page 135. But, as stated on that page we combine the three middle elements together to obtain a triple. That is the quintuple a,b,c,d,e becomes the triple a, $\sqrt{(b^2+c^2+d^2)}$,e.

$$\begin{array}{lccc} & 8 & 4 & 1 & 9 \\ & 6 & 3 & 2 & 7 \quad - \\ \hline \hat{AOB}: & 8\times6+4\times3+1\times2, & \sqrt{(4\times2-1\times3)^2+(1\times6-8\times2)^2+(8\times3-4\times6)^2}, & 9\times7 \\ \hline = & 62 & \sqrt{125} & 63 \end{array}$$

We combine the 3 middle elements by squaring them, adding and taking the square root.
So the triple 62, $\sqrt{125}$, 63 represents angle AOB (see proof on Page 161).
Note that the middle element of the triple can be calculated more easily from the other elements than by the method shown above.
That is $63^2 - 62^2 = 125$, (or $(63 + 62)(63 - 62) = 125$ in this case).

Example 2 Find a triple for the angle between the planes $2x + y - 2z = 5$ and $3x - 6y - 2z = 4$.

The angle between the planes is the same as the angle between their normals. Therefore we need angle POQ where P is P(2,1,–2) and Q is Q(3,–6,–2):

$$\begin{array}{l|cccc} P & 2 & 1 & -2 & 3 \\ Q & 3 & -6 & -2 & 7 \quad - \\ \hline \text{angle POQ} & 4 & - & 21 & \end{array}$$

which is the required triple.

Example 3 Find the angle between the lines $\frac{x}{1}=\frac{y-1}{2}=\frac{z-2}{3}$ and $\frac{x}{2}=\frac{y+1}{3}=\frac{z+2}{4}$.

As in Example 6 in Chapter 5 we can transpose the origin to the point of intersection of the lines and find the angle between $\frac{x}{1}=\frac{y}{2}=\frac{z}{3}$ and $\frac{x}{2}=\frac{y}{3}=\frac{z}{4}$:

$$\begin{array}{cccc} 1 & 2 & 3 & \sqrt{14} \\ 2 & 3 & 4 & \sqrt{29} \quad - \\ \hline 20 & - & & \sqrt{29\times14} \end{array}$$

is the required triple.

Example 4 Find the distance of the point (1,–2, 3) from the plane $3x - 2y + z + 1 = 0$.

Since if $x = y = 0$, $z = -1$, the plane meets the z-axis at $z = -1$.
We therefore transpose the origin to this point (i.e. to (0,0,–1)) and find the distance of P(1, –2, 4) from $3x - 2y + z = 0$:

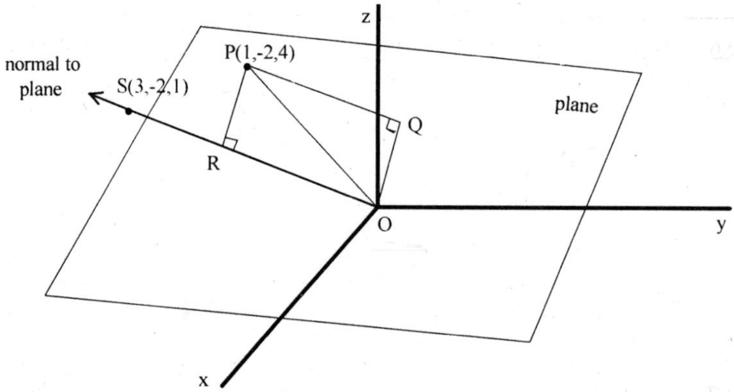

We want PQ and PQ = RO. Since we know P(1,–2,4) and S(3,–2,1) we can find the triple for ΔPOR:

P	1	–2	4	$\sqrt{21}$
R	3	–2	1	$\sqrt{14}$
$P\hat{O}R$	11		-	$\sqrt{14 \times 21}$

$\therefore$ $PQ = \frac{11}{\sqrt{14}}$ (as OP= $\sqrt{21}$).

Example 5 Find the perpendicular distance of the point (2, 1, –3) from the line

$$\frac{x-1}{2} = \frac{y-2}{-1} = \frac{z}{-1}.$$

We know (1, 2, 0) lies on the line and we transpose the origin to that point and then find the distance of P(1, –1, –3) from $\frac{x}{2} = \frac{y}{-1} = \frac{z}{-1}$.

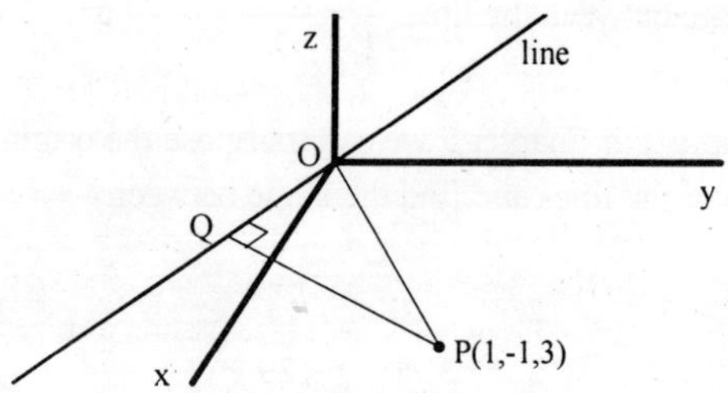

1	–1	–3	$\sqrt{11}$
2	–1	–1	$\sqrt{6}$
6	$\sqrt{30}$	$\sqrt{66}$	

Since OP = $\sqrt{11}$ we divide through by $\sqrt{6}$.

$\therefore$ $PQ = \sqrt{5}$.

Example 6 If A(1, 7, 3) and B(2, 1, –3) show that OA and OB are perpendicular.

The triple for 90° is 0, 1, 1, so in finding angle AOB we require that the vertical products forming the first number of the triple should add up to zero:

1	7	3	-
2	1	–3	-
(2+7–9)	-	-	

which they do.

∴ OA and OB are perpendicular.

Example 7 Find the equation of the straight line joining (2, 0, 5) and (1, 2, 3).

We eliminate first the z-dimension and retain x and y, by considering the projection of the line on the x,y plane.
Then using the method described on Page 48 the equation of the line joining (2,0) and (1,2) is $y = -2x + 4$.
Then eliminating the x-dimension, the equation of the line joining (0,5) and (2,3) in the y,z plane is $-2z = 2y - 10$. That is, $y = 5 - z$.

Then bringing the equations $y = -2x + 4$ and $y = 5 - z$ together the equations of the line in 3-dimensions are: $-2x + 4 = y = 5 - z$.

It is most convenient to write the given coordinates one under the other for this:

2	0	5
1	2	3

WORK AND MOMENT

Now that quadruple subtraction has been defined it is possible to find the work done and the moment of a force in the 3-dimensional case.

Example 8 Find the work done when the force $\mathbf{F} = 2\mathbf{i} + 2\mathbf{j} + \mathbf{k}$ causes a displacement $\mathbf{r} = 6\mathbf{i} + 3\mathbf{j} + 2\mathbf{k}$. Find also the magnitude of the moment about the origin of **F** when it acts through the point (6,3,2). The units are Newtons and metres.

We use the method of quadruple subtraction described above:

F	2	2	1	3	
r	6	3	2	7	–
		20	$\sqrt{41}$	21	

The work done is $2 \times 6 + 2 \times 3 + 1 \times 2 = 20$ Nm.
The third element of the triple is $3 \times 7 = 21$.
The magnitude of the moment is the middle element of the triple, $\sqrt{41}$ Nm.

Thus we see that there is no need to evaluate a 3 × 3 determinant when finding the magnitude of a moment in 3-dimensions, as the middle element is most easily found from the other two elements of the triple.

The scalar and vector products of vector algebra are neatly unified in the following result:

$$\mathbf{F} - \mathbf{r} = \mathbf{F}\cdot\mathbf{r},\ |\mathbf{F}\times\mathbf{r}|,\ |\mathbf{F}||\mathbf{r}|$$

in which $\mathbf{F} - \mathbf{r}$ indicates quadruple subtraction (as carried out in Example 1 above) of the elements of $\mathbf{F}$ and $\mathbf{r}$.

Finally it is worth noting that if a vector cross-product $\mathbf{F}\times\mathbf{r}$ is required, rather than just the magnitude, this is available in the original definition of quadruple subtraction. If the middle three elements of the quintuple, which combine to form the middle element of the triple, are found they give the three elements of the vector product though the outer elements must be reversed and their signs changed.

Example 9 If $\mathbf{F} = 2\mathbf{i} + 2\mathbf{j} + \mathbf{k}$ and $\mathbf{r} = 6\mathbf{i} + 3\mathbf{j} + 2\mathbf{k}$ find $\mathbf{F}\times\mathbf{r}$.

The original quadruple subtraction gives:

F	2	2	1		3	
r	6	3	2		7	–
F–r	20	6	2	–1	21	

The middle three elements give $\mathbf{F}\times\mathbf{r} = \mathbf{i} + 2\mathbf{j} - 6\mathbf{k}$.

ROTATION ABOUT A COORDINATE AXIS

If we are rotating a point about, say, the x-axis, the x-coordinate of the point will not change and the rotation is in a plane parallel to the y,z plane.

Example 10 Rotate the point P(4,2,3) about the x-axis by the angle 4,3,5.

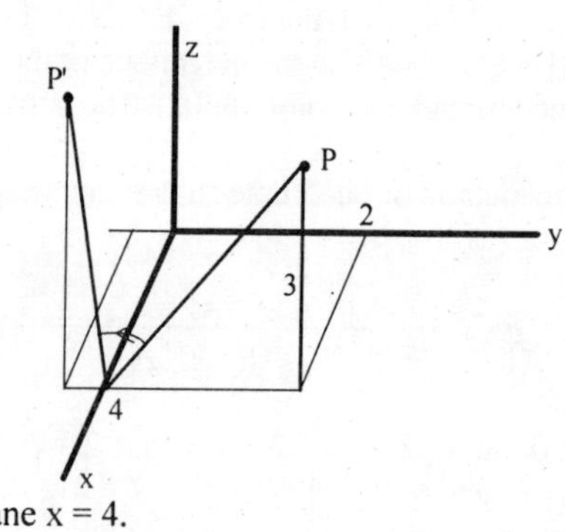

We rotate in the plane x = 4.

In this plane we are adding a 4,3,5 triangle to a 2,3,- triangle:

4	2	3	$\sqrt{13}$	
	4	3	5	+
	$-\frac{1}{5}$	$\frac{18}{5}$	$\sqrt{13}$	

The coordinates of the rotated point P' are therefore (4, –0.2, 3.6).

Example 11 Rotate P(4,2,3) by angle 4,3,5 about a) the y-axis b) the z-axis.

a) The y coordinate of P is unchanged.

4	2	3	-	
4		3	5	+
$\frac{7}{5}$	2	$\frac{24}{5}$		

$\therefore$ P'(1.4, 2, 4.8).

b) The z-coordinate is unchanged.

4	2	3	-	
4	3		5	+
$\frac{10}{5}$	$\frac{20}{5}$	3		

$\therefore$ P'(2, 4, 3).

We do not need to calculate the hypotenuse for P each time and since we invariably divide through by the third element of the rotation triple we can disregard the 4th column and write the answer straight down.

Note that for rotations about the y-axis, adding the triples rotates clockwise from the positive y-direction (but anticlockwise for the other two axes).

Combining rotations like this enables us to perform rotational transformations (as shown in section 2 of the next chapter).

We can also use this method to rotate shapes in 3-dimensional space (about an inclined axis) by making small rotations about two or three of the coordinate axes in turn.

3-DIMENSIONAL ROTATION OF CURVES

The method of rotating lines and curves outlined in Chapter 4 can now be extended.

Example 12 Rotate the curve $\mathbf{y = x^2, z = 0}$ by the angle in the triple **4,3,5** about the x-axis.

This is the curve $y = x^2$ in 3-dimensional space: without the qualification, $z = 0$, the equation $y = x^2$ would represent a surface.

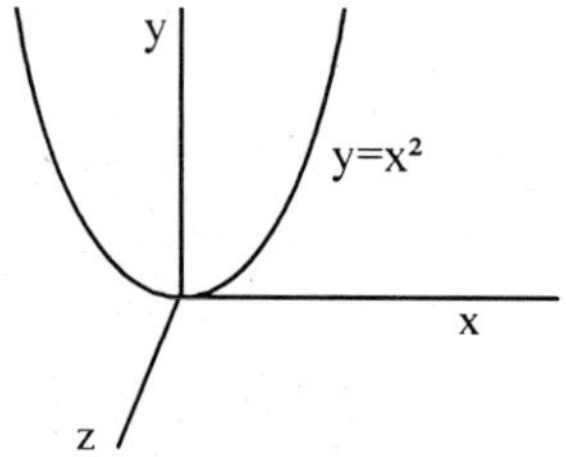

Rotation about the x-axis means it is in the yz plane so we get the transformational equations as follows:

X	Y	Z	r	
	4	3	5	–
x	y	z	r	

using the same notation as in Chapter 4.
This gives $x = X,\ y = \frac{4Y+3Z}{5},\ z = \frac{4Z-3Y}{5}$.

Substituting these into $y = x^2$ and $z = 0$ gives $\frac{4Y+3Z}{5} = x^2$ and $\frac{4Z-3Y}{5} = 0$.

So that **$5x^2 - 4y - 3z = 0,\ 3y = 4z$**.

Example 13 Similarly to rotate **$y = x^2$, $z=0$** about the y-axis we have:

X	Y	Z	r	
4		3	5	–
$x = \frac{4X+3Z}{5}$,	$y = Y$,	$z = \frac{4Z-3X}{5}$,	r	

So that $y = x^2$, $z = 0$ becomes **$25y = (4x + 3z)^2,\ 3x = 4z$**.

Other curves and surfaces can be rotated in a similar way.

ROTATION

Let us consider what we mean by rotations in 3-dimensions. Suppose that we have a centre of rotation and another point which has to be rotated. Then if the rotation has to be in a plane the centre of rotation and the point must both lie in that plane. We can let the origin of coordinates be at the centre of rotation, because if it is not there we can easily transpose the origin to that point. So to perform a particular rotation we need the coordinates of the point to be rotated, the angle of the rotation and the plane of the rotation: which is defined by the direction of its normal.

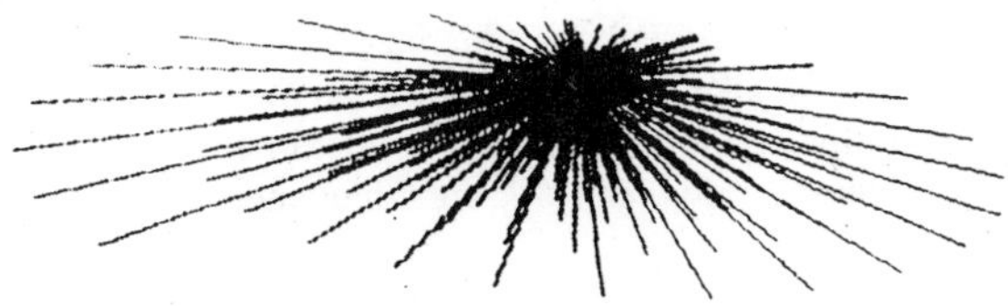

These pictures were produced by a BBC microcomputer using a program that worked entirely with triples and quadruples.
Not a single square root or trigonometrical function was evaluated in the execution of the program.

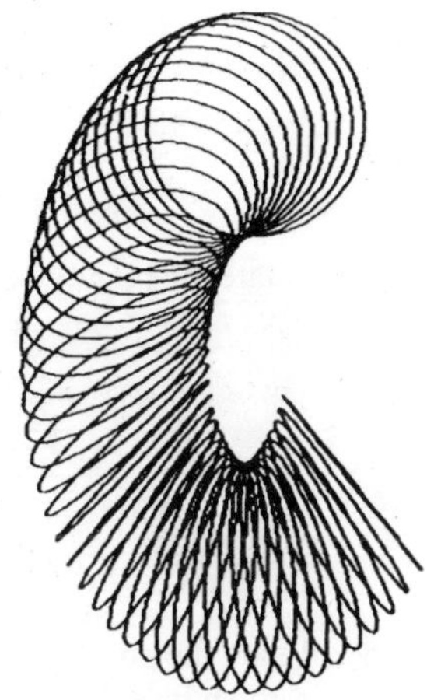

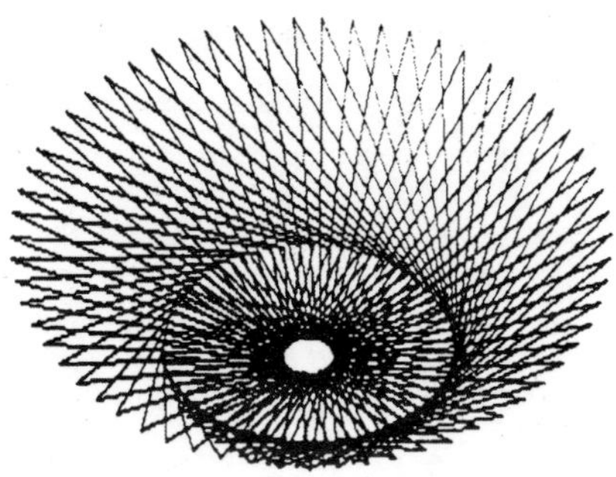

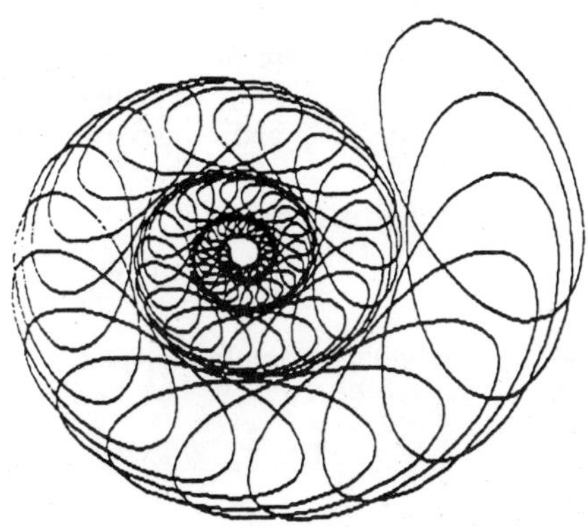

Example 14 Rotate the point P(9, 4, –5) through an angle of 90° in the plane whose normal is given by the quadruple 2,3,6,7.

Since the line OP lies in the plane, and the normal is perpendicular to it, the normal is perpendicular to OP. This is confirmed (see Example 6, Page 139) by the total of the three vertical products below: $2\times9 + 3\times4 + 6\times-5 = 0$.

$$\begin{array}{ccc} 2 & 3 & 6 \\ 9 & 4 & -5 \\ \hline 3\times-5 - 6\times4 & 6\times9 - 2\times-5 & 2\times4 - 3\times9 \end{array} \quad = \quad -39,\ \ 64,\ \ -19$$

So the rotated point is $P'\frac{1}{7}(-39, 64, -19)$.

That is, for the x-coordinate we take the cross-product in the 2nd and 3rd columns, for y we take the cross-product (in the reverse direction) in the 1st and 3rd columns and for z the cross-product in the 1st and 2nd columns. All of these must be divided by 7, the 4th element of the quadruple.

In general for 90° rotations, if the point is p,q,r and the quadruple defining the normal direction is a,b,c,R, then the rotated point has coordinates:

$$\begin{array}{ccc} a & b & c \\ p & q & r \\ \hline \frac{1}{R}(br-cq, & cp-ar, & aq-bp) \end{array}$$

Example 15 Rotate the point P(-6,1,2) through 90° in the plane with normal 2,-10,11,15.

$$\begin{array}{cccc} 2 & -10 & 11 & 15 \\ -6 & 1 & 2 & \\ \hline -31 & -70 & -58 & \end{array}$$

Therefore $P' = \frac{1}{15}(-31, -70, -58)$.

Example 16 Rotate the point P(3, 4, –3) through the angle 4,3,5 in the plane whose normal is given by 2,3,6,7.

Given a normal:	a	b	c	R
a point:	p	q	r	
and a rotation:	l	m	n	

then P' is given by:

$$\frac{1}{nR}\left(m(br-cq)+plR,\ m(cp-ar)+qlR,\ m(aq-bp)+rlR\right)$$

So in this example we have:

$$\begin{array}{cccc} 2 & 3 & 6 & 7 \\ 3 & 4 & -3 & \\ 4 & 3 & 5 & \\ \hline \frac{1}{35}(3\times-33+3\times28, & 3\times24+4\times28, & 3\times-1+-3\times28) & \end{array}$$

$\therefore\ P' = \frac{1}{35}(-15, 184, -87)$.

CONICOIDS

In Chapter 8 we saw how triples can be used to plot points on ellipses and hyperbolas. This idea can be extended to 3-dimensional conics, called conicoids, by using quadruples instead of triples.

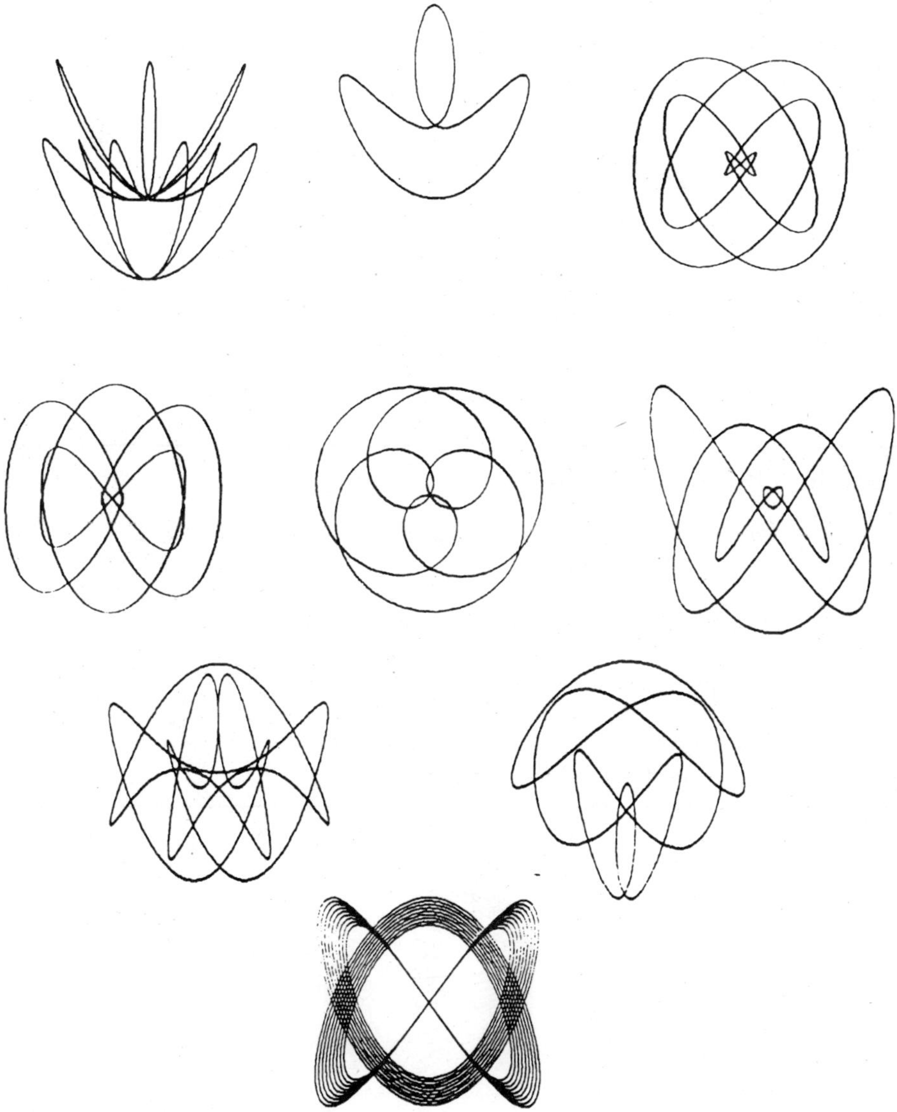

These Lissajous figures were generated using only perfect triples, no trig. functions.

<u>Example 17</u> The quadruple **2x, 3y, 4z, 12** represents an ellipsoid whose centre is at the origin and whose axes are the coordinate axes. We can find points on the ellipsoid in the same way as for the 2-dimensional case; we select a quadruple of our choice and cross-multiply:

2x	**3y**	**4z**	**12**
20	**12**	**9**	**25**

$x = \frac{12\times20}{2\times25} = 4.8, \quad y = \frac{12\times12}{3\times25} = 1.92, \quad z = \frac{12\times9}{4\times25} = 1.08.$

Similarly using the quadruple 2,2,1,3 we get $(4, \frac{8}{3}, 1)$,
and using 1,0,0,1 we get (6, 0, 0).

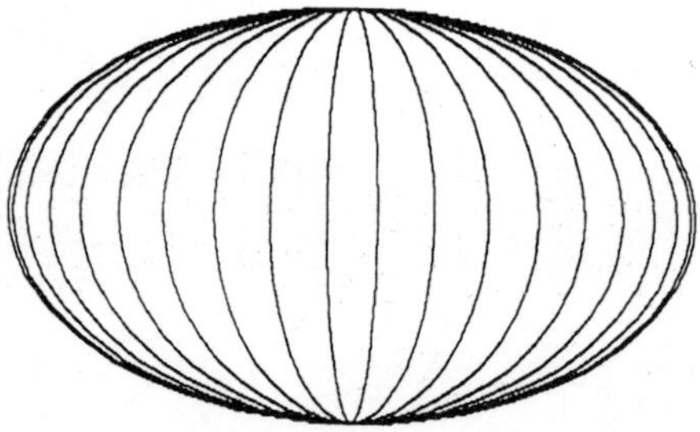

An ellipsoid

<u>Example 18</u> Plot the hyperboloid **12, z, 2y, x**.

We can fit the quadruple 16,15,12,25:

12	**z**	**2y**	**x**
16	**15**	**12**	**25**

We proceed as we did in the 2-dimensional case:

$x = \frac{12\times25}{1\times16} = 18.75, \qquad y = \frac{12\times12}{2\times16} = 4.5, \qquad z = \frac{12\times15}{1\times16} = 11.25.$

So (18.75, 4.5, 11.25) lies on the hyperboloid.
And so on.

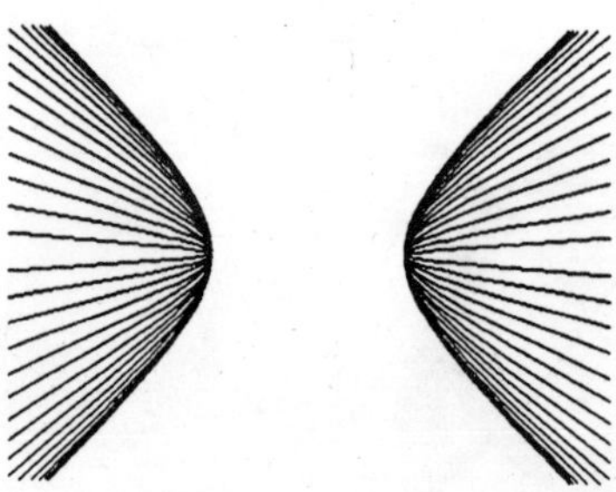

A hyperboloid of two sheets

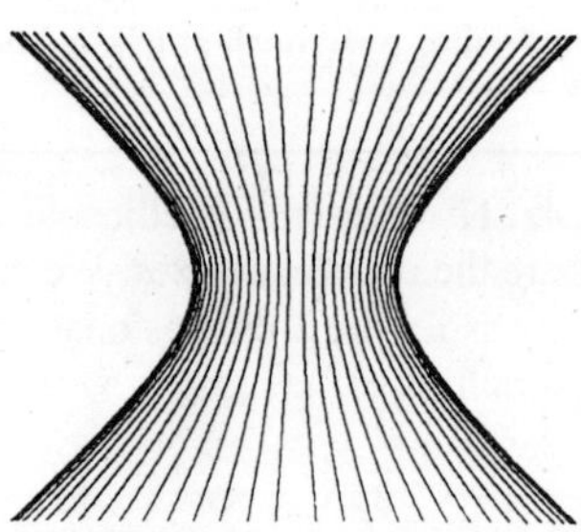

A hyperboloid of one sheet

16 QUADRUPLES IN ASTRONOMY

1. ADDITION OF PERPENDICULAR TRIPLES

In astronomy positions in space are often described by an angle measured from some reference direction in a reference plane, and another angle perpendicular to this plane:

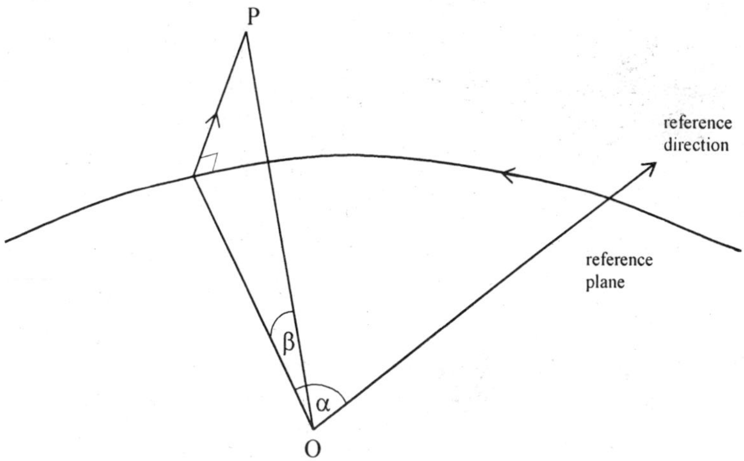

So the position of P is described by two perpendicular angles α, β.
But since a position in space can also be described by a quadruple, there must be a relationship between the elements of the triples for α, β and the corresponding quadruple for the same position.

Suppose that the two triples are **α)3 4 5** **β)8 15 17** then the equivalent quadruple is **24 32 75 85.**
For 24 and 32 we multiply the first two elements of the first triple by the first element of the second triple,
and for 75 and 85 we multiply the last two elements of the last triple by the last element of the first triple.

This is an application of the Vedic formula *The First by the First and the Last by the Last.* We can see why this procedure works if we place the 8 side of the β triple along OQ below, which is the hypotenuse of the α triple:

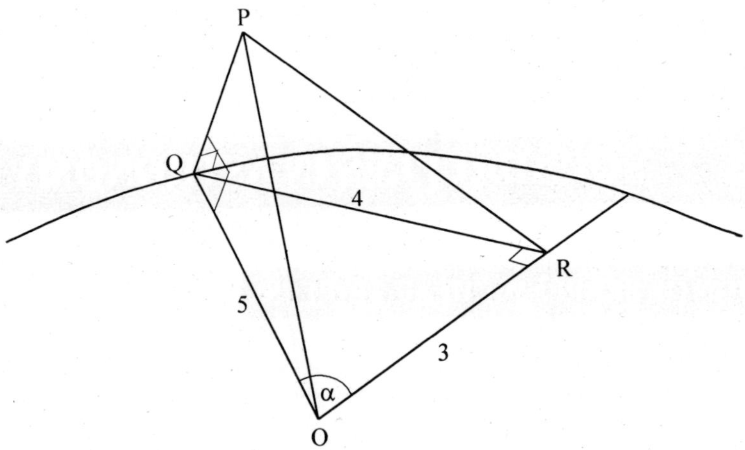

Using *Proportion* we take the β triple as $5, \frac{5\times15}{8}, \frac{5\times17}{8}$ to make the two triples "fit".
Then we can see that the quadruple is $3, 4, \frac{5\times15}{8}, \frac{5\times17}{8} = 24, 32, 75, 85$.

Thus two perfect triples produce a perfect quadruple.

Another useful way of adding two triples to produce a quadruple is discussed in Section 5.

<u>Example 1</u> The equatorial coordinates (α,δ) of a star are given by the triples
α) 4 3 5 δ) 12 5 13. Find the star's rectangular equatorial coordinates.

This is exactly similar to the problem discussed above:
the answer is 48, 36, 25, 65.

It is also easy to do the reverse operation: find the two triples from the quadruple.
Given the quadruple 48, 36, 25, 65 from Example 1 we can put 48 and 36 as the first two elements of the a triple, cancel by 12 and then insert the last element, giving α)4 3 5;
and we can put 25 and 65 as the last two elements of the d triple, cancel by 5 and insert the first element, getting δ)12 5 13.

2. CHANGE OF COORDINATE SYSTEM

Here we use the method from the last chapter for rotating about a coordinate axis.
If we have the coordinates of a point P(x', y', z') in a rectangular coordinate system x', y', z' and we want its coordinates in another rectangular coordinate system x, y, z with the same origin, we can get them by using the triple rotation method described in Examples 10, 11 in the previous chapter.

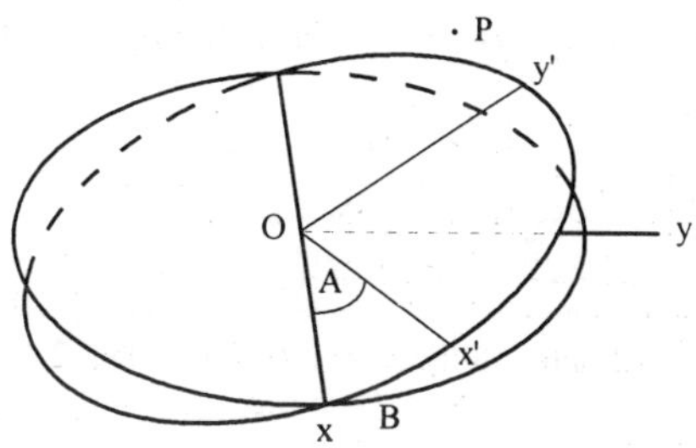

Suppose that the x-y plane of these two systems are inclined at B)l, m, n, and let us suppose for the moment that Ox is along the line of intersection of these planes, as shown above.

Let the angle between Ox and Ox' be given by A)i, j, k.
So A and B describe the orientation of the two frames of reference and we suppose that these, and P(x', y', z') are known. We want P(x, y, z).

Consider the xx'y' plane viewed from the z'O direction:

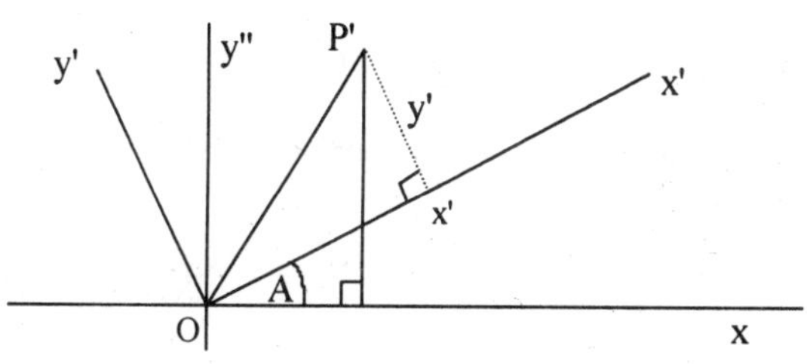

P' is the projection of P on this plane.
We are effectively rotating x'Oy' about the Oz' axis by the angle A so that Ox' coincides with Ox and Oy' falls on Oy".

	x'	y'	z'	r	
A	i	j	k		+
	x'i-y'j	y'i+x'j	z'k	rk	= a, b, c, d say.

Next we rotate this axis system xy''z' about the x-axis by the angle B:

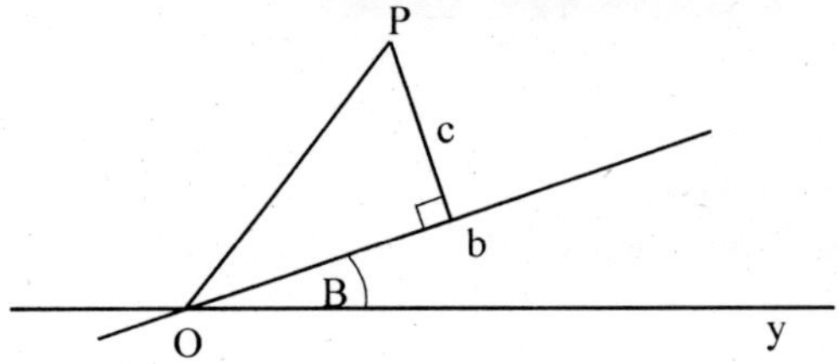

Note that a corner of the triangle drawn above is not at O but at a point a on the x-axis.
Adding we get:

	a	b	c	d	
B		l	m	n	+
	an	nl-cm	cl+bm	dn	= A, B, C, D say.

<u>Example 2</u> Find the coordinates in the x, y, z frame of P(2,2,1) given A)4 3 5 and B)12 5 13.

	2	2	1	3	
A	4	3	5		+
	2	14	5	15	
B		12	5	13	+
	26	143	130	195	
=	2	11	10	15	
=	0.4	2.2	2	3	

So P has coordinates <u>(0.4, 2.2, 2)</u> in the other frame.

If however the Ox axis is not along the line of intersection of the xy and x'y' planes a further rotation is needed: about Oz.
Suppose C)p,q,s is the angle between Ox and the line of intersection of the planes:

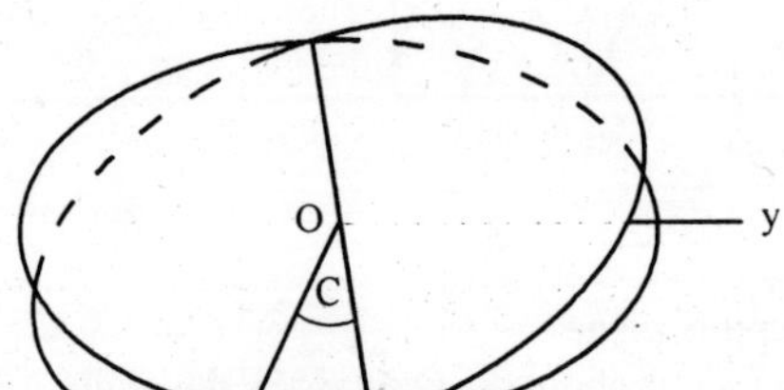

P will be given by:

A	B	C	D	
p	q	s		+
Ap-Bq	Bp+Aq	cs	Ds	

Clearly this method could be extended to 3 or more inclined planes.
Note also that the quadruple for P could represent a third orbit plane rather than the position of a point.

3. QUADRUPLES AND ORBITS

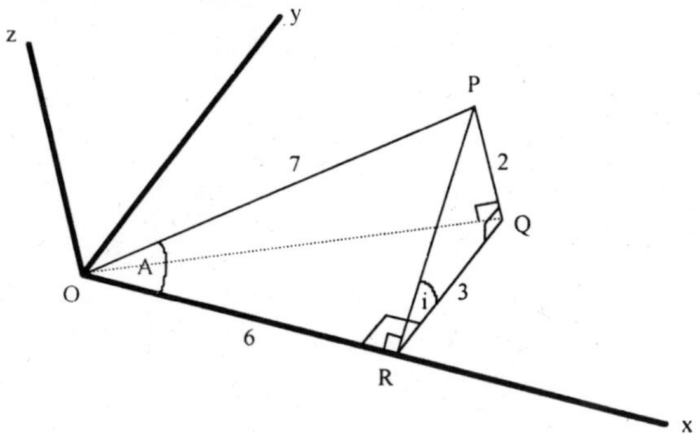

The diagram shows the quadruple 6, 3, 2, 7.
Angle POR = A, i.e. $A = \cos^{-1}\left(\frac{6}{7}\right)$.
Angle PRQ = i, i.e. $i = \tan^{-1}\left(\frac{2}{3}\right)$.

A quadruple can be used to represent an orbit crossing, in this case, the x-axis.
i is the inclination of the orbit to the reference frame and the angle A defines a particular position in the orbit.
Sections 4 to 13 in this chapter form a sequence, exploring the ramifications of this use of quadruples and developing the arithmetic for combining quadruples and their code numbers.

4. QUADRUPLE FOR GIVEN i AND A

It is sometimes necessary to know the quadruple which has a given inclination, i, and argument, A. Suppose, for example, that we have i) 12 5 13 and A) 4 3 5 and that we want to know the quadruple that they define:

$$\begin{array}{cccc} & \mathbf{12} & \mathbf{5} & \mathbf{13} \\ & \mathbf{4} & \mathbf{3} & \mathbf{5} \\ \hline \mathbf{52} & \mathbf{36} & \mathbf{15} & \mathbf{65} \\ \hline \end{array}$$

We use the *Vertically and Crosswise* formula:

for 52 and 36 we multiply crosswise, 4 × 13 and 12 × 3,
and for 15 and 65 we multiply vertically, 5 × 3 and 13 × 5.

This result is easily proved by using a method similar to that used in Section 1 for the addition of two perpendicular triples.

5. INCLINATION OF ORBIT

In the quadruple 6, 3, 2, 7 the inclination, i, is determined by the two inner elements 3, 2 and is independent of the values of the outer elements, which determine A (see Section 1).

Consider the general quadruple

$$c^2-d^2+e^2,\ 2cd,\ 2de,\ c^2+d^2+e^2$$

generated by the code numbers c, d, e.

The ratio of the inner elements of the quadruple is the same as the ratio of the outer elements of the code numbers (c:e).
So all quadruples generated by, say, 3, d, 2 will be of the form x, 3, 2, r for all values of d.

But as we have just seen it is the inner pair of elements in a quadruple that determine the inclination of the orbit. It follows that all quadruples generated by 3, d, 2 will be in the same orbit plane (defined by POR in the previous diagram).

And in general all quadruples generated by c, d, e where c and e are constants will be inclined at $\tan^{-1}\left(\frac{e}{c}\right)$ to the xy plane for all d. We will take this further in Section 10.

6. QUADRUPLE SUBTRACTION

So far we have only defined quadruple subtraction (Page 135), the result being a quintuple. We would like to be able to add and subtract quadruples and for the result to be a quadruple. And, of course, we would like the process to serve some useful purpose.

Let us impose the condition (to hold in the following sections) that when adding or subtracting two quadruples the ratio of their inner elements is the same. The geometrical consequence of this, as discussed in the last section, is that the two quadruples will be in the same orbit. Furthermore this will result in one of the elements of the quintuple being zero: thereby giving a quadruple rather than a quintuple.

For example, suppose we want to find 2,6,3,7 – 2,2,1,3 in which the ratio of the inner elements is the same, i.e. 6:3 = 2:1. The subtraction gives (see Page 135):

$$\begin{array}{lllll} \mathbf{2} & \mathbf{6} & \mathbf{3} & & \mathbf{7} \\ \mathbf{2} & \mathbf{2} & \mathbf{1} & & \mathbf{3} \\ \hline \mathbf{19} & \mathbf{8} & \mathbf{4} & \mathbf{0} & \mathbf{21} \end{array} \; - \qquad (2)$$

which has a zero element and so can be compressed into the quadruple 19, 8, 4, 21.

Note that the inner elements of this quadruple are in the same ratio as the two subtracted quadruples: 8:4 = 2:1, so it is in the same plane also. And note the operation of the Vedic formula *One in Ratio, the Other One Zero.*

7. QUADRUPLE ADDITION

Before we look further at the geometrical significance of this subtraction method let us consider addition. Can we, for example, add 19, 8, 4, 21 and 2, 2, 1,3 and get 2, 6, 3, 7 (see (2) above)?

For subtraction we have:

$$\begin{array}{cccc} x & y & z & r \\ X & Y & Z & R \\ \hline xX + yY + zZ, & yX - xY, & zX - xZ, & rR \end{array} \quad - \qquad (3)$$

For addition we will have:

$$\begin{array}{cccc} x & y & z & r \\ X & Y & Z & R \\ \hline xX - yY - zZ, & yX + xY, & zX + xZ, & rR \end{array} \quad + \qquad (4)$$

in which the patterns are all the same and some signs are reversed.

That this has the desired effect of reversing the subtraction can be verified by taking the subtraction:

$$\begin{array}{cccc} x & y & z & r \\ X & ny & nz & R \\ \hline xX + ny^2 + nz^2, & yX - nxy, & zX - nxz, & rR \end{array} \quad -$$

and adding the last two lines as indicated in (4).
On cancelling and dividing through by $X^2+n^2y^2+n^2z^2$ ($=R^2$) the result should be x,y,z,r thus verifying that (4) is a legitimate addition process.

So let us now add 19,8,4,21 and 2,2,1,3 using this addition algorithm:

$$\begin{array}{cccc} \mathbf{19} & \mathbf{8} & \mathbf{4} & \mathbf{21} \\ \mathbf{2} & \mathbf{2} & \mathbf{1} & \mathbf{3} \\ \hline \mathbf{18} & \mathbf{54} & \mathbf{27} & \mathbf{63} \end{array} \begin{array}{c} \\ + \\ = \underline{\mathbf{2,6,3,7}} \end{array}$$

Example 3

$$\begin{array}{cccc} \mathbf{1} & \mathbf{2} & \mathbf{2} & \mathbf{3} \\ \mathbf{1} & \mathbf{2} & \mathbf{2} & \mathbf{3} \\ \hline \mathbf{-7} & \mathbf{4} & \mathbf{4} & \mathbf{9} \end{array} \begin{array}{c} \\ + \\ \\ \end{array} \quad \text{and} \quad \begin{array}{cccc} \mathbf{-7} & \mathbf{4} & \mathbf{4} & \mathbf{9} \\ \mathbf{1} & \mathbf{2} & \mathbf{2} & \mathbf{3} \\ \hline \mathbf{9} & \mathbf{18} & \mathbf{18} & \mathbf{27} \end{array} \begin{array}{c} \\ - \\ = \underline{\mathbf{1,2,2,3}}. \end{array}$$

8. DOUBLING AND HALVING A QUADRUPLE

Doubling a quadruple follows directly from the last section. For equal quadruples the addition process given in (4) yields:

$$2(x,y,z,r) = x^2 - y^2 - z^2, \ 2xy, \ 2xz, \ r^2.$$

So 2(7,4,4,9) = 17, 56, 56, 81.

The formula for halving a quadruple is:

$$\tfrac{1}{2}(x,y,z,r) = x + r,\ y,\ z,\ (\sqrt{r}) \quad \text{or} \quad -(x + r),\ -y,\ -z,\ (\sqrt{r}).$$

The first three elements define the result. $\sqrt{r}$ should only be inserted for the fourth element when x,y,z,r is a perfect quadruple and the first three elements have been cancelled down.

So $\tfrac{1}{2}(17,56,56,81) = 98,\ 56,\ 56,\ (\sqrt{81}) = 7, 4, 4, 9$ or $-7, -4, -4, 9$.

The geometrical proof of this halving process follows very similar lines to that for halving the angle in a triple. (see Chapter 2).

9. CODE NUMBER ADDITION AND SUBTRACTION

For triples it was found to be a great advantage to be able to add and subtract the triple code numbers rather than the triples themselves, because we often work with the code numbers or require the code number of a final result rather than the triple itself. It is the same with the quadruples, we would like to be able to add and subtract the code numbers as well as the quadruples.

Let c,d,e and c,D,e be the code numbers of two quadruples which we want to add (we can use c,D,e rather than nc,D,ne because we can divide code numbers through by any constant without changing the quadruple it generates).

If we obtain the quadruples that these code numbers generate (as shown on Page 134), add the quadruples (as shown by (4) on Page 153), obtain the code numbers of the result (as shown on Page 134), simplify and divide through by 2(d + D) we obtain:

$$\begin{array}{ccc} & \mathbf{d} & \\ \mathbf{c} & & \mathbf{e} \\ & \mathbf{D} & \\ \hline \mathbf{c(c^2+e^2-dD),} & \mathbf{(c^2+e^2)(d+D),} & \mathbf{e(c^2+e^2-dD)} \end{array} \quad +$$

for addition and:

$$\begin{array}{ccc} & \mathbf{d} & \\ \mathbf{c} & & \mathbf{e} \\ & \mathbf{D} & \\ \hline \mathbf{c(c^2+e^2+dD),} & \mathbf{(c^2+e^2)(d-D),} & \mathbf{e(c^2+e^2+dD)} \end{array} \quad -$$

for subtraction.

Note that because c and e are the same in both code numbers we need write them only once, so that the numbers form a diamond pattern.

Example 4 Find the code numbers of the quadruple which is the sum of the quadruples whose code numbers are 3,1,2 and 3,3,2.

Using the above formula:

	1		
3		2	+
	3		
30	**52**	**20**	= **15, 26, 10.**

We can of course also do this the long way: find the quadruples from the code numbers, add them and find the code numbers of the result. The quadruples generated by 3,1,2 and 3,3,2 are 6,3,2,7 and 2,9,6,11 and their sum is –27,60,40,77. The code numbers of this quadruple are 15,26,10, as before.

10. ANGLE IN A QUADRUPLE

The angle in a quadruple is the angle A in the diagram on Page 151.
We now ask, what is the relationship between the angles in two quadruples and the angle in their sum or difference?

In fact the addition and subtraction processes defined here have the effect of adding and subtracting the angles respectively.

For example, let the angles in the quadruples in (2) be A, B, C.

A	2	6	3	7	
B	2	2	1	3	–
C	19	8	4	21	

Then A – B = C, or $\cos^{-1}(\frac{2}{7}) - \cos^{-1}(\frac{2}{3}) = \cos^{-1}(\frac{19}{21})$.

We can easily verify this by triple subtraction:

$\cos^{-1}(\frac{2}{7})$	2	$\sqrt{45}$	7	
$\cos^{-1}(\frac{2}{3})$	2	$\sqrt{5}$	3	–
	$4+\sqrt{45\times5}$,	-,	21	

and the angle in 19, - , 21 is $\cos^{-1}(\frac{19}{21})$

Let us now add the quadruples 1,2,2,3 and 7,4,4,9:

D	**1**	**2**	**2**	**3**	
E	**7**	**4**	**4**	**9**	+
F	**–9**	**18**	**18**	**27**	= **–1,2,2,3.**

And adding the angles in D and E:

$\cos^{-1}(\frac{1}{3})$	1	$\sqrt{8}$	3	
$\cos^{-1}(\frac{7}{9})$	7	$\sqrt{32}$	9	+
	$7-\sqrt{8\times 32}$,	-,	27	

and the angle in -1, - , 3 is $\cos^{-1}(-\frac{1}{3})$.

So D + E = F, and adding the quadruples adds the angles in the quadruples (see proof at the end of the Appendix).

It appears then that this addition and subtraction method for quadruples would have useful astronomical applications: when, for example, a body in an inclined orbit (inclined to some reference plane) advances in its orbit by a certain amount, and we want to know its new position relative to the same reference plane.

11. ANGULAR ADVANCE

Now we know (Section 5) that quadruples generated by code numbers c,d,e, where c and e are given, are in the same plane for all values of d. For example, consider the code numbers 2,d,1 for various values of d:

code numbers			generated quadruple			
2	**0**	**1**	**1**	**0**	**0**	**1**
2	**1**	**1**	**2**	**2**	**1**	**3**
2	**2**	**1**	**1**	**8**	**4**	**9**
2	**3**	**1**	**–2**	**6**	**3**	**7**
2	**4**	**1**	**–11**	**16**	**8**	**21**
etc.			etc.			

Note: 1) whole number values of d have been selected here, but any values may be chosen,
2) as d increases, the angle in the generated quadruple increases,
3) the first element of the 4th and 5th quadruples is negative, indicating that the angles are greater than 90°.

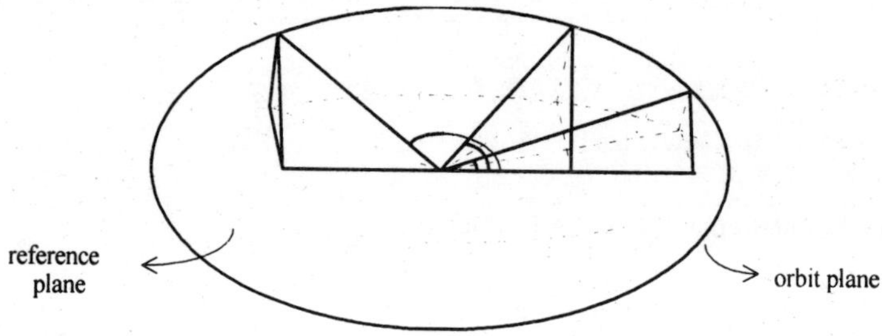

The diagram above shows three quadruples with increasing angles and their relationship to the reference plane.

In fact as d increases from 0 to ∞ (infinity), A increases from 0° to 180°.
Quadruples with arguments greater than 180° are produced by code numbers with negative d values: as d increases from $-\infty$ to 0, A increases from 180° to 360°.

The angular advance as d increases from 0 to 1, from 1 to 2 etc. can be found by subtracting the quadruples:

2	2	1	3	
1	0	0	1	–
2	2	1	3	

1	8	4	9	
2	2	1	3	–
22	14	7	27	

The angular advance in these two cases is $\cos^{-1}(\frac{2}{3})$ and $\cos^{-1}(\frac{22}{27})$.
It is worth noting that the code numbers of 22,14,7,27 are 14,5,7, so that the d number of this quadruple is $\frac{5}{7}$.

Now we may want to add a quadruple to another quadruple so that the added quadruple increases the angle by a certain given value.

12. RELATIONSHIP BETWEEN d AND A

This relationship is expressed by $\mathbf{c^2 + e^2 = (kd)^2}$, where **k** is the triple code number of A.
So c, e, kd is a triple.

This is easily proved as follows. Code numbers c,d,e generate the quadruple $c^2-d^2+e^2, 2cd, 2de, c^2+d^2+e^2$. This contains the angle given by the triple $c^2-d^2+e^2$, - , $c^2+d^2+e^2$.

The code number of this triple (see Page 57, Example 8) is given by $k = \sqrt{\dfrac{(c^2+e^2)}{d^2}} = \dfrac{\sqrt{c^2+e^2}}{d}$.

So $\mathbf{c^2 + e^2 = (kd)^2}$.

<u>Example 5</u> Advance from 12,4,3,13 by the angle in the triple whose code number is 10.

We have k=10 and c=4, e=3.
The above formula gives the d number for the quadruple which is to be added to the given quadruple: $4^2+3^2 = (10d)^2$, from which d=½.
This means that the quadruple to be added has code numbers 4,½,3 = 8,1,6.
This quadruple is 99, 16, 12, 101:

12	4	3	13	
99	16	12	101	+
1088	588	441	1313	

So 1088, 588, 441, 1313 contains an angle which is the sum of the original angle and the angle in the triple with code number 10.

Alternatively we can do the addition using code numbers.
In Section 9 we saw how to add code numbers instead of the quadruples themselves.
In the present example we obtained d=½, giving 8,1,6 as the code numbers of the quadruple to be added.
If we now add the code numbers:

$$\begin{array}{ccccc} & 1 & & & \\ 4 & & 3 & + & \\ & \tfrac{1}{2} & & & \\ \hline \underline{98} & \underline{37\tfrac{1}{2}} & \underline{73\tfrac{1}{2}} & = & \underline{196,\ 75,\ 147}. \end{array}$$

or, alternatively:

$$\begin{array}{ccccc} & 2 & & & \\ 8 & & 6 & + & \\ & 1 & & & \\ \hline \underline{98\times 8} & \underline{100\times 3} & \underline{98\times 6} & = & \underline{196,\ 75,\ 147}. \end{array}$$

And the code numbers of 1088, 588, 441, 1313 are 196, 75, 147.

<u>Example 6</u> Given that the Sun was at the position given by the quadruple whose code numbers are 23,5,10 (which it was on 1978 April 13th), find its rectangular coordinates when it has moved through 43°36' (which was on 1978 May 28th).

The Sun's "orbit" is inclined at about 23°27' to the reference (equatorial) plane. This orbit is called the ecliptic. All quadruples generated from the code numbers 23,d,10 are in the ecliptic. We consider in the next section the problem of finding the values of c and e for a particular given orbit inclination and we also derive the code numbers 23,d,10 for the inclination of the ecliptic.

Now we require the code number of 43°36'. In Chapter 9 we went into details of how to find the code number of a given angle and so we simply state here that it is k=2.5.

Next we find the value of d: $d = \frac{\sqrt{23^2+10^2}}{2.5} \approx 10$.
So we add the code numbers 23,5,10 and 23,10,10:

$$\begin{array}{ccccc} & 5 & & & \\ 23 & & 10 & + & \\ & 10 & & & \\ \hline \underline{579\times 23} & \underline{629\times 15} & \underline{5790} & = & \underline{4439,\ 3145,\ 1930} \end{array}$$ which

are the code numbers of the final position.

Finally, if required, we can generate the quadruple from these code numbers. This is not difficult using the Vedic formula *Vertically and Crosswise,* and working from left to right (see Reference 8 Pages 27, 41, 68) we can easily obtain the final coordinates to 4 figures; these are 1354, 2791, 1214.

The angle between this direction and the correct one is 18'. We have of course rounded off in three places as well as using the approximate code numbers 23,d,10, but the method is exact and so a more accurate answer is obtainable by keeping more significant figures.

13. TO OBTAIN A QUADRUPLE WITH A GIVEN INCLINATION

In the example above we used the code numbers 23,d,10 to represent all quadruples in the plane of the ecliptic, relative to a rectangular, equatorial reference system. These numbers once obtained can be used whenever calculations involving this plane are to be made. Standard planes* whose code numbers would be worth listing would include:

1. inclination of the ecliptic to the equator,
2. inclination of the equator to a particular horizon,
3. inclination of the various planetary orbits to the ecliptic.

We need a way of obtaining these code numbers to any desired degree of accuracy.

In fact the inclination described by the code numbers 23,d,10 is $\tan^{-1}(\frac{10}{23})$ which is 23°30' and not 23°27'. The following method can be used to obtain the code numbers for any inclination to any desired degree of accuracy.

* There are small or slow variations in the inclinations in some cases.

We will take the inclination 23°27' as our example and generate code numbers for this to various degrees of accuracy. It will be seen from the geometry of the problem that we want two integers whose ratio is tan23°27' (see Section 6). We therefore begin by expressing tan 23°27' as a continued fraction (see Reference 10, Pages 114-5). We get:

$$\tan 23°27' = \frac{1}{2+}\,\frac{1}{3+}\,\frac{1}{3+}\,\frac{1}{1+}\,\frac{1}{1+}\,\frac{1}{1+}\,\frac{1}{3+}\,.$$

This gives the convergents: $\frac{1}{2}\;\frac{3}{7}\;\frac{10}{23}\;\frac{13}{30}\;\frac{23}{53}\;\frac{36}{83}\;\frac{131}{302}\,.$

These fractions give the desired ratio to increasing degrees of accuracy. The third of these, $\frac{10}{23}$, is the ratio used in Example 6. Had we used the sixth, $\frac{36}{83}$, we would have been working with an angle of 23°27'. So any quadruple generated by 83,d,36 will be in an orbit inclined at 23°27' (to the nearest minute of arc).

APPENDIX: Proofs etc.

SUM OF TWO TRIPLES

Let the triples be A)x, y, z and B)X, Y, Z and let them be added as shown:

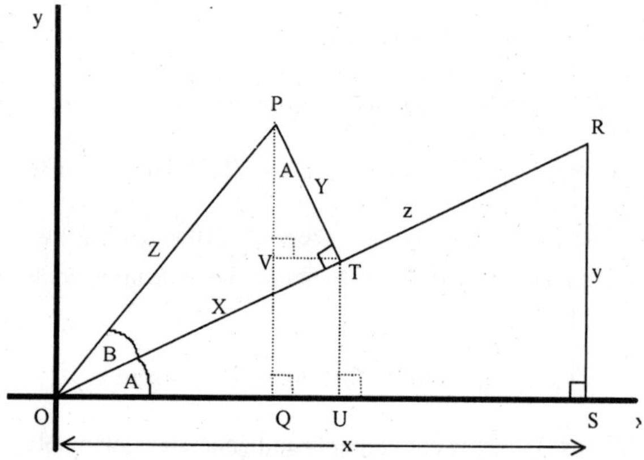

$V\hat{P}T$=A as $V\hat{T}O$=A and $V\hat{T}P$=90°–A.
∴ triangles PVT, OSR and OUT are similar, with sides in the ratio Y:z:X.
We require OQ and PQ:
OQ=OU – QU = OU – VT= $\frac{xX}{z} - \frac{yY}{z} = \frac{xX-yY}{z}$. And PQ=QV + VP = UT + VP= $\frac{xX}{z} + \frac{yY}{z} = \frac{xX+yY}{z}$

∴ the base, height and hypotenuse of triangle OQP are: $\frac{xX-yY}{z}$, $\frac{yX+xY}{z}$, Z.
Then multiplying the triangle lengths by z we get:

A :	x	y	z	
B :	X	Y	Z	+
A+B:	xX–yY	yX+xY	zZ	

EQUATION OF A LINE

For the points (a,b) and (c,d) the Vedic method gives the equation of the line through them as:

a	b
c	d

(a-c)y = (b-d)x + D (1)

where D = $\begin{vmatrix} a & b \\ c & d \end{vmatrix}$.

Since the gradient of (1) is clearly $\frac{b-d}{a-c}$ as required we have only to show that D is satisfied by (a,b) or (c,d).

Putting (a,b) into (1): (a-c)b = (b-d)a + D.

So $D = (a-c)b - (b-d)a = \begin{vmatrix} a-c & b-d \\ a & b \end{vmatrix} = \begin{vmatrix} a-c & b-d \\ c & d \end{vmatrix} = \begin{vmatrix} a & b \\ c & d \end{vmatrix}$ as required by the Vedic method.

ANGLE BETWEEN TWO LINES

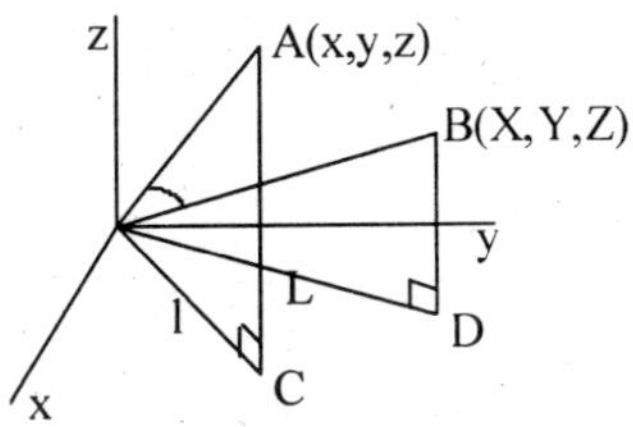

We require a triple for $A\hat{O}B$.
Consider the xy-plane:

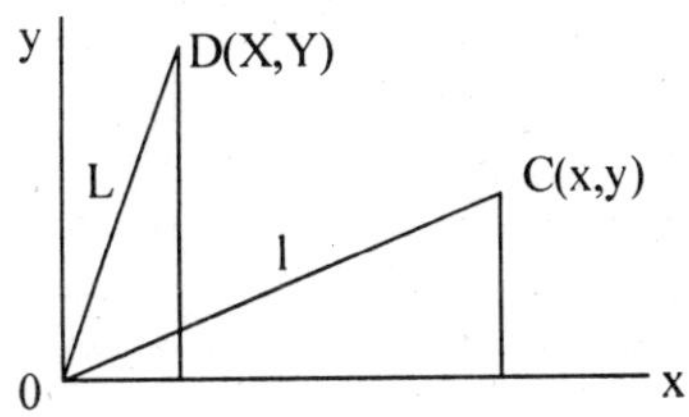

Rotate A, B, C and D clockwise about the z-axis by angle x, y,

∴ for D':

D :	X	Y	L
C :	x	y	l –
D' :	$\frac{xX+yY}{l}$,	$\frac{Yx-Xy}{l}$,	L

Then we have:

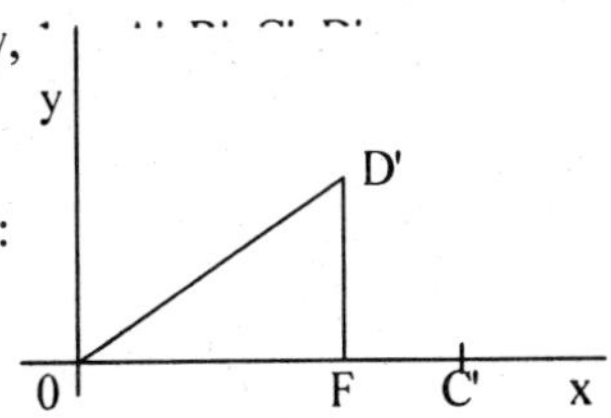

Consider now the xz-plane: drop a perpendicular from B' to meet the xz-plane at B" and let OB"=p. Then for $A'\hat{O}B''$:

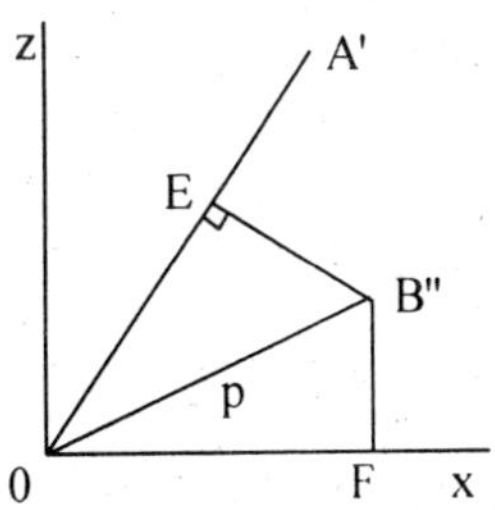

A' :	1	z	r
B" :	(xX+yY)/l	Z	p –
$A'\hat{O}B''$:	xX + yY + zZ,	-,	rp

∴ OE = $\frac{xX+yY+zZ}{r}$ and OB' = R ∴ $A'\hat{O}B'$) $\frac{xX+yY+zZ}{r}$, -, R = $A\hat{O}B$

∴ $A\hat{O}B$) xX+yY+zZ, -, rR.

Then the third element of this triple is given by:

$(EB')^2 = (x^2+y^2+z^2)(X^2+Y^2+Z^2) - (xX+yY+zZ)^2$

which becomes $(EB')^2 = (xY - yX)^2 + (xZ - zX)^2 + (yZ - zY)^2$.

$\therefore\ A\hat{O}B)$ (xX+yY+zZ), $\sqrt{(xY-yX)^2+(xZ-zX)^2+(yZ-zY)^2}$, Rr

IF X^2+1 DIVIDES X+Y, IT DIVIDES XY–1

Let two triples have code numbers x, y (both natural numbers), then the triple sum of x and y is xy–1,x+y.

We want to show that if (x^2+1) divides x+y exactly then it also divides xy–1 exactly.

Let $x+y = n(x^2+1)$ (1)
then $y = nx^2 + n - x$.
$\therefore\ xy-1 = x(nx^2 + n - x) - 1 = nx^3 + nx - x^2 - 1$
$\therefore\ xy-1 = (x^2 + 1)(nx - 1)$ (2)
$\therefore\ x^2+1$ divides xy–1.

Also the triple sum of x and y is xy–1, $x+y = (x^2+1)(nx-1), n(x^2+1)$ from (1) and (2) above
= nx–1, n
so that nx–1 can be obtained immediately once n is known (see Example 18, Page 99).
Similarly, the triple difference of x and y is xy+1, y–x and it can be shown in the same way that if (x^2+1) divides y–x then it also divides xy+1 and that xy+1, y–x = nx+1, n.

THE CODE-NUMBER FORMULA

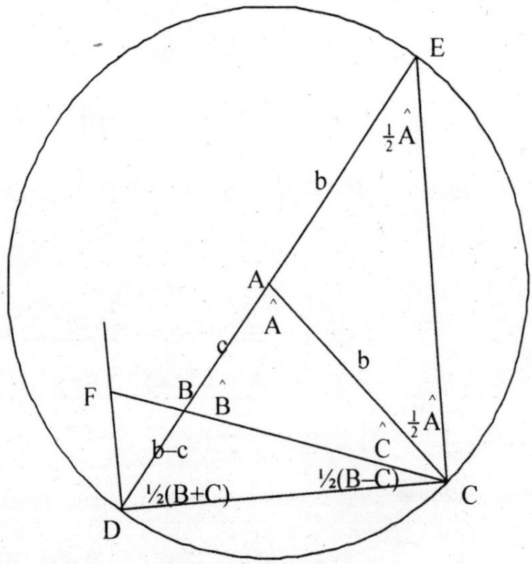

With A as centre draw a circle, radius b, around triangle ABC.
Extend AB in both directions to complete the diameter ED.
Join EC and DC.
Draw DF parallel to CE and let F be on CB produced.

Then BE=b+c, BD=b-c and $A\hat{C}E = A\hat{E}C = \frac{1}{2}A$

$E\hat{C}D = C\hat{D}F = 90°$, and since A+B+C=180°, $F\hat{C}D = \frac{1}{2}(B-C)$.

Also $E\hat{A}C = B + C$, so that $E\hat{D}C = \frac{1}{2}(B+C)$.

Now triangles ECB and DFB are similar so $\frac{CE}{FD} = \frac{b+c}{b-c}$ or $\frac{CD}{FD} \div \frac{CD}{CE} = \frac{b+c}{b-c}$.

But $\frac{CD}{FD}$ = code numbers of (B–C) and $\frac{CD}{CE}$ = code numbers of (B+C).

Therefore $\frac{c(B-C)}{c(B+C)} = \frac{b+c}{b-c}$ where c(B–C) and c(B+C) mean the code numbers of (B–C) and the code numbers of (B+C) respectively.

ACCURACY OF CODE NUMBERS WHEN CALCULATING ANGLES

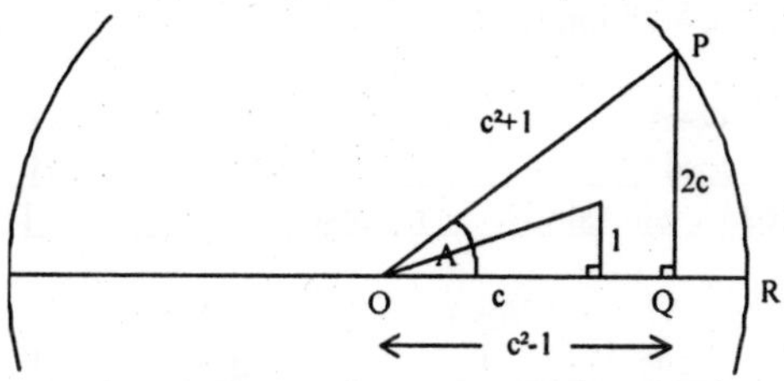

This diagram shows the triple generated from the code number c.
From this we see that the exact algebraical relationship between c and A is c=cot½A. (1)
In the conversion of code numbers into angles, if we assume that the known angles (from the table) are available to any degree of accuracy so that the error is negligible, the only inaccuracy arising is due to the use of the approximation $cA \approx 2$.
From equation (1) above we get

$$A = 2\tan^{-1}\left(\tfrac{1}{c}\right)$$

and using the series expansion for $\tan^{-1}x$:

$$A = \frac{2}{c} - \frac{2}{3c^3} + \frac{2}{5c^5} - \frac{2}{7c^7} \dots$$

Neglecting terms with c^3 and higher powers we have $A \approx \frac{2}{c}$ or $cA \approx 2$ as derived on Page 96.
But since the next term, in the series for A above, is $-\frac{2}{3c^3}$, this can be used either for a more precise value of A, or as an estimate of the inaccuracy in using $\frac{2}{c}$ for A.
For accuracy to 4 decimal places we require $\frac{2}{3c^3}$<0.00005, from which $c \geq 24$ (as given on Page 91).

When converting angles into code numbers there are two points at which we approximate: in using $cA \approx 2$, and in finding the second number and using an approximation to it.
Considering $cA \approx 2$ first, from (1) using the expansion for cotx we get:

$$c = \frac{2}{A} - \frac{A}{6} - \frac{A^3}{360} \dots$$

so that $\frac{A}{6}$ is a measure of the inaccuracy of c in using $\frac{2}{A}$ (or it can be used to obtain a better approximation to c).
However if we assume the use of code numbers that are integers or halves the greatest inaccuracy that can arise is 0.00002 radians (for A=0.702). And since the inaccuracy is usually much smaller we will not consider this any further.

The relationship between a small change, δc, in c and the corresponding change, δA, in A, is found by differentiating (1) above:
c=cot½A so $\frac{dc}{dA}=-\frac{1}{2}\operatorname{cosec}^2\frac{1}{2}A=-\frac{1}{2}\left(1+\cot^2\frac{1}{2}A\right)=-\frac{1}{2}\left(1+c^2\right)$.
Therefore $\delta A=\frac{-2\delta c}{1+c^2}$ and $\delta c=-\frac{1}{2}\left(1+c^2\right)\delta A$

Now looking at the effect of using an approximation to the second code number, having obtained its value from $cA\approx 2$:
if A is an angle, c its exact code number, and (c+x) the code number actually used, then x is the difference between the two and the inaccuracy in A due to using (c+x) is $\frac{2}{c}-\frac{2}{c+x}=\frac{2x}{c(c+x)}\approx\frac{2x}{c^2}$.

ACCURACY OF SIN, COS AND TAN

Using the triple notation and triple addition we have:

A:	cosA	sinA	1
dA:	cosdA	sindA	1 +
A+dA:	cosAcosdA−sinAsindA	sinAcosdA+cosAsindA	1

Or using the approximation:

A:	cosA	sinA	1
dA:	1	dA	1 +
A+dA:	cosA−dAsinA	sinA+dAcosA	1

Considering sine, we are approximating sinAcosdA+cosAsindA with sinA+dAcosA.

$$\text{Now } \sin A\cos dA+\cos A\sin dA=\sin A\left(1-\frac{(dA)^2}{2!}+\ldots\right)+\cos A\left(dA+\ldots\right)$$
$$=\sin A+dA\cos A-\frac{(dA)^2}{2}\sin A\ldots$$

The first two terms here give the approximation used, and the third term is an error term. Since the maximum absolute value of sinA is 1 we get, for 4-figure accuracy:
$\frac{(dA)^2}{2}<0.00005$, which gives dA<0.01.

We get the same result in the case of triple subtraction. For cosine the result is similar: $\frac{(dA)^2}{2}\cos A$ and for tangent we get $\frac{(dA)^3}{3}\sec^2 A$, though a little more algebra is needed.

INVERSE SINE AND INVERSE COSINE

We want A+dA where A is the angle in a known triple and dA=x−A, x being the required angle.
Since $\frac{d\sin A}{dA}=\cos A$, $\delta A\approx\frac{\delta\sin A}{\cos A}$ and using the same notation as on Page 104:
$x=A+\delta A\approx A+\frac{\delta\sin A}{\cos A}=A+\left(\frac{m}{n}-\frac{q}{r}\right)\times\frac{r}{p}=A+\frac{mr-nq}{pn}$.

Similarly from $\frac{d\cos A}{dA}=-\sin A$, we can get: $x\approx A-\frac{lr-pn}{nq}$.

ACCURACY OF INVERSE SINE AND INVERSE COSINE

How close should the standard perfect triple be to the given triple to ensure, say, 4-figure accuracy?

The error involved in using the approximation $\delta A \approx \frac{\delta \sin A}{\cos A}$ is $\delta A - \frac{\delta \sin A}{\cos A} = \delta A - \frac{\sin(A+\delta A) - \sin A}{\cos A}$

$$= \delta A - \frac{\sin A \cos \delta A + \cos A \sin \delta A - \sin A}{\cos A}$$

$$= \delta A - \frac{\sin A\left(1 - \frac{(\delta A)^2}{2!} + \ldots\right) + \cos A\left(\delta A - \frac{(\delta A)^3}{3!} + \ldots\right) - \sin A}{\cos A}$$

$$= \frac{(\delta A)^2}{2}\tan A, \text{ neglecting terms in } (\delta A)^3 \text{ and higher.}$$

So for $A < \frac{\pi}{4}$, for 4-figure accuracy we require $\frac{(\delta A)^2}{2} < 0.00005$ or $\delta A < 0.01$.

QUADRUPLE ADDITION ADDS THE ANGLES IN THE QUADRUPLES

Quadruple addition is defined as:

x	y	z	r	
X	Y	Z	R	+
xX – yY – zZ,	yX + xY,	zX + xZ,	rR	

In which $x^2 + y^2 + z^2 = r^2$ and $X^2 + Y^2 + Z^2 = R^2$, (1)
and these quadruples have the property that yZ = zY. (2)

We have to show that $\cos^{-1}\left(\frac{x}{r}\right) + \cos^{-1}\left(\frac{X}{R}\right) = \cos^{-1}\left(\frac{xX - yY - zZ}{rR}\right)$.

In triples we need:	x,	$\sqrt{r^2 - x^2}$,	r	
	X,	$\sqrt{R^2 - X^2}$,	R	+
	xX-yY-zZ,	-,	rR	

That is, we must show that: $xX - \sqrt{r^2 - x^2}\sqrt{R^2 - X^2} = xX - yY - zZ$.

But, from (1) the left-hand side of this is $xX - \sqrt{y^2 + z^2}\sqrt{Y^2 + Z^2}$

$= xX - \sqrt{(yY)^2 + (zZ)^2 + (yZ)^2 + (zY)^2}$

$= xX - \sqrt{(yY)^2 + (zZ)^2 + 2yYzZ}$ using (2) above

$= xX - yY - zZ$ as required.

ANSWERS

CHAPTER 2

Exercise A

1. 13, 84, 85 2. 44, 117, 125 3. 133, 156, 205 4. 21, 220, 221 5. –13, 84, 85
6. 7, 24, 25 7. 161, 240, 289 8. 41, 840, 841 9. –119, 120, 169 10. –161, 240, 289
11. 117, 44, 125 12. 63, 16, 65 13. 323, 36, 325 14. 56, –33, 65 15. –7,–24, 25
16. 120,–119, 169 17. –33, 56, 65 18. –323,–36, 325 19. 836, 123, 845 20. 169, 0, 169 (= 1,0,1)

Exercise B

1. 5,–12, 13 2. 4, –3, 5 3. –15, –8, 17 4. 1, 7, $5\sqrt{2}$
5a 5, –12, 13 b 17, –7, $13\sqrt{2}$ c –120, –119, 169 d –120, 119, 169
6. $1-\sqrt{3}, 1+\sqrt{3}, 2\sqrt{2}$ 7. $-1-\sqrt{3}, -1+\sqrt{3}, 2\sqrt{2}$ 8. $\sqrt{3}+1, 1-\sqrt{3}, 2\sqrt{2}$
9. $-1+\sqrt{3}, -\sqrt{3}-1, 2\sqrt{2}$ 10. –1, –1, $\sqrt{2}$ 12. 29,3,- (i.e. 3,5,- – 3,4,5)

Exercise C

1. 3, 2, $\sqrt{13}$ 2. 4, 1, $\sqrt{17}$ 3. 5, 3, $\sqrt{34}$ 4. 12, 5, 13 5. 1, 3, $\sqrt{10}$
6. 5, 12, 13 7. 3, 4, 5 8. –7, 1, $\sqrt{50}$ 9a 4, 3, 5 b 24, 7, 25
c 2, 1, $\sqrt{5}$ d 2, 11, $5\sqrt{5}$ e –527, –336, 625 f –117, 44, 125

CHAPTER 3

Exercise A

1. $\frac{63}{65}$ 2. $-\frac{65}{33}$ 3. $-\frac{24}{7}$ 4. $\frac{7}{17}$ 5. $\frac{2}{5}$ 6. $-\frac{140}{171}$ 7. $-\frac{828}{2197}$
8. $\frac{1}{\sqrt{26}}$ 9. $-\frac{5}{\sqrt{26}}$ 10. $\frac{2\sqrt{34}}{5+3\sqrt{3}}$

Exercise B

1. $\frac{\pi}{4}$ 15. 1, $\frac{\pi}{8}$ 16. 0,±1,1; 1,1,$\sqrt{2}$; –1, –1, $\sqrt{2}$ 17. $\pm\sqrt{3}, -1, 2$; 0, 1, 1
18. 1, 0, 1; $\pm\sqrt{2}, \pm 1, \sqrt{3}$ (4 solutions) 19, $\pm\sqrt{3}, \pm 1, 2$ 20. $\frac{\sqrt{2}}{10}$

Exercise C

1. 24, –7, 25; 4, –3, 5 2. 0, 1, 1; 12, 5, 13 3. 5, 12, 13; 3, 4, 5
4. 45n° 5. (45+60n)°; (135+180n)° 6. (±45+180n)°; (±60+360n)° 7. $1+\frac{\sqrt{3}}{2}$

CHAPTER 4

Exercise A

1. (–5,3) 2. (7,–2) 3. (4,13) 4. (–2,9) 5. (–0.4,2.2) 6. (–3.6,0.2)
7. (1.8,–2.4) 8. (4,0) 9. (–2.2,4.6) 10. $(-2\sqrt{3}-3, 3\sqrt{3}-2)$ 11. 62, 25, -

Exercise B

1. (–4,7) 2. (–3,8) 3. (2,2) 4. (1.4,4.8) 5. (8.16,10.12)
6. (2,–11) 7. (–5,4) 8. (2.8,4.6) 9. –2.2,4.6) 10. (–6.6,–4.8)

CHAPTER 5

Exercise A

1. 3.6 2. 3.84 3. 1.8 4. 5.1 5. $\frac{20}{13}$ 6. $\frac{13}{\sqrt{10}}$
7. 1.6 8. $\frac{6}{\sqrt{5}}$

Exercise B

1. 7, 1, $\sqrt{50}$ 2. 18, 1, $5\sqrt{13}$ 3. 5, 1, $\sqrt{26}$ 4. 7, 1, $\sqrt{50}$ 5. 24, 7, 25
6. 63, 16, 65 7. 45° 8. 12, 5, 13 9. 90°
10. 7y = 4x (find 31,27 – 3,1 to get a triple for the angle between the lines; halve this triple to get 5,1 and add 3,1 to get 7,4)

Exercise C

1. 5y = 2x + 1 2. 3y = 2x – 19 3. 4y = 3x + 31 4. 2y = 3x + 9
5. y = x + 7 6. y = –x – 3

CHAPTER 6

Exercise A

1. 29 2. 42 3. 79 4. 64
5. 1023, 64, 1025 6. 684, 37, 685 7. 783, 56, 785 8. 5100, 101, 5101

Exercise B

1. 99, 20, 101 2. 399, 40, 401 3. 9999, 200, 10001 4. 55, 48, 73 5. 7, 24, 25
6. 77, 36, 85 7. 20, 21, 29 8. 105, 88, 137 9. 60, 91, 109 10. 20, 99, 101
11. 63, 16, 65 12. 20, 21, 29 13. –3, –4, 5 14. 5, –12, 13 15. –4, 3, 5
16. 7, 5 17. 9, 5 18. 7, 2 19. 19, 15 20. 17, 3
21. 13, 7 22. 7, 9 23. 11, –2 24. 4, –7 25. 20, –9

Exercise C

1. 11, 10 2. 51, 43 3. 25, 19 4. 78, 11 5. 14, –5 6. 5, 3 7. 8, 5
8. 13, 3 9. 10, 1 10. 11, 5 11. 4, –3 12. 3, 5 13. 7, –2 14. –3, 17
15. 1, 5 16. 7, 1 17. –1, 2 18. 1 19. 1 20. 2 21. 2
22. 4 23. 1 24. 4 25. 3 26. 3 27. 0° 28. 0°
29. 180° 30. 90° 31. 270° 32. 180°

CHAPTER 7

Exercise A

1. BD = 5, DC = 1, AF = $\frac{19}{7}$, FB = $\frac{30}{7}$, CE = $\frac{6}{5}$, EA = $\frac{19}{5}$
2. A)19, -, 35; B)5, -, 7; C)1, -, 5 3. C)–1, -, 15
4. 41 5. 53 6. 37 7. $\sqrt{32.8}$ 8. AP = 1.4 cm

Exercise B

1. 78 2. -, 11, 15 3. 21 4. $28-3\sqrt{51}, 4\sqrt{51}+21, 50$ 5. 21.25 6. 7

Exercise C

1. B)7, 30, -; C)25, 18, - 2. B)1, 4, -; C)13, 16, - 3. B)–7, 23, -; C)1, 1, -
4. B)29, 7, -; C)149, 7, - 5. B)120, 191, -; C)160, 159, -

CHAPTER 9

Exercise B

1. 0.0488 2. 0.0290 3. 0.0392 4. 0.0286 5. 0.0142
6. 80 7. 45 8. 100 9. 250 10. 140

Exercise C

1. 1.0808 2. 0.3597 3. 0.8364 4. 0.8532 5. 0.9879
6. 0.6058 7. 0.4442 8. 0.8732 9. 0.5970 10. 0.4061

11. 0.6018 12. 0.4570 13. 0.8097 14. 0.5829 15. 1.0384

Exercise D
1. 0.7812 2. 1.1386 3. 0.7497 4. 1.0997 5. 0.8419
6. 0.6272 7. 0.4765 8. 0.3082 9. 0.9365 10. 0.0659
11. $\frac{\pi}{2}$–0.1993=1.3715 12. $\frac{\pi}{2}-\frac{2}{45}$=1.5264 13. 1.176+$\frac{2}{135}$=1.1908
14. $\pi-\left(0.3948-\frac{2}{83}\right)=2.7709$ 15. $-\left(\pi-\left(0.3948+\frac{2}{99}\right)\right)=-2.7266$

Exercise E
1. 128, 41 2. 64, 23 3. 227, 61 4. 67, 36 5. 127, 28
6. 133, 46 7. 121, 47 8. 147, 41 9. 215, 58 10. 79, 42
11. 91, 43 12. 111, 53 13. 139, 43 14. 47, 12

Exercise F
1. 70, 23 a=0.0001 2. 29, 10 a=0.0001 3. 119, 60 4. 117, 29 a=0.0001 5. 109, 6
6. 74, 3 7. 49, 8 a=0.0003 8. 79, 26 a=0.0001 9. 14, 3 a=0.0002 10. 47, 4 a=0.0002
11. 29, 7 a=0.0003 12. 23, 10 a=0.0003 13. 41, 16 a=0.0001 14. 27, 8 a=0.0001 15. 26, 7
16. 91, 24 a=0.0003 17. 53, 5 a=0.0001 18a $\sqrt{3}$+2,1 b 1+$\sqrt{2}$,1 c 3, $\sqrt{3}=\sqrt{3}$,1

CHAPTER 10
Exercise A
1. 11.974, 5.0624, 13 2. 35.2436, 11.2895, 37 3. 3.9805, 3.026, 5
4. 21.62, 19.349, 29 5. 44.9048, 28.153, 53 (using 45, 28, 53)

Exercise B
1. 0.3303+$\frac{1}{105}$=0.3398 2. 0.6435–$\frac{1}{28}$=0.608 3. $\frac{\pi}{2}-\left(0.49-\frac{1}{225}\right)=1.0852$
4. 0.3948+$\frac{1}{216}$=0.3994 5. $\frac{\pi}{2}-0.2838-\frac{1}{168}=1.2810$ 6. 1.5708–0.3303+$\frac{1}{12\times19}$=1.2449
7. ½(0.49)=0.2450 8. 1.5708–0.2450=1.3258 9. 0.32175
10. ½(1.5708–0.3948)=0.5880 11. ½(0.6435–$\frac{2}{43}$)=0.2985 12. ½(0.9273–$\frac{2}{17}$)=0.405

CHAPTER 12
Exercise A
1. $\pm3\sqrt{3}$; -9 2. $\pm\sqrt{3}$; $\pm9\sqrt{3}$ 3. 0.4636; 0.32175

Exercise B
1a $40\sqrt{5}$; 160, 100, - b 19, 2, - c 173, 8, -

2a	0	–10	-	0	–10	-
	$10\sqrt{3}$	10–10t	-	10	$10\sqrt{3}-10t$	-
	$10\sqrt{3}\,t$	10t–5t²	-	10t	$10\sqrt{3}\,t-5t^2$	-

b $\left(20\sqrt{3},0\right)$; $\left(20,20\sqrt{3}-20\right)$
3. (24,13); 16m

Exercise C
1. $\frac{50}{\sqrt{3}}$; $\frac{100}{\sqrt{3}}$ 2. $\frac{40}{\sqrt{3}}$ 3. $\frac{210}{24}$ (top); $\frac{25\times30}{24}$ (bottom) 4. $\frac{36}{5}$; $\frac{6}{5}\sqrt{61}$

REFERENCES

1. Tirthaji B. K. Vedic Mathematics. Delhi: Motilal Banarsidass, 1965.

2. Thom A., A. S. Thom. Megalithic Remains in Britain and Brittany. Oxford: The Clarendon Press, 1978.

3. Wood J. E. Sun Moon and Standing Stones. Oxford University Press, 1978.

4. Amma T. A. Geometry in Ancient and Medieval India. Delhi: Motilal Banarsidass, 1979.

5. Neugebauer O. The Exact Sciences in Antiquity. New York: Dover, 1969.

6. de Solla Price D. J. The Babylonian "Pythagorean Triangle" Tablet. Centaurus, 10, pages 1-13, 1964.

7. Thibaut G. On the Sulvasutras. Journal, Asiatic Society of Bengal, Part 1, 1875.

8. Williams K. R. Discover Vedic Mathematics. Motilal Banarsidass, 1984.

9. Nicholas A. P., K. Williams, J. Pickles. Vertically and Crosswise. Motilal Banarsidass, 1984.

10. Williams K. R. Astronomical Applications of Vedic Mathematics. Motilal Banarsidass, 2000.

11. Maor E. e: The Story of a Number Princeton University Press, 1994.

INDEX OF VEDIC FORMULAE

INDEX

INDEX